Becoming a Researcher:

Making the Transition to Graduate School

Educational Contributions in Ethnobiology

Educational Contributions in Ethnobiology

Marsha Quinlan and Justin Nolan, Series Editors

Educational Contributions in Ethnobiology is a peer-reviewed monograph series dedicated to original book-length publications in ethnobiology and related social and natural sciences. The volumes, rather than present research, present methods, theory, and pedagogy that advances ethnobiological research.

Becoming a Researcher: Making the Transition to Graduate School
Steve Wolverton

Becoming a Researcher:
Making the Transition to Graduate School

Steve Wolverton

Society of Ethnobiology
2018

Library of Congress Control Number: 2017954203

ISBN 978-0-9990759-0-6 (paperback)
ISBN 978-0-9990759-1-3 (PDF)

Society of Ethnobiology
Department of Sociology & Anthropology, University of Puget Sound
1500 North Warner St., CMB#1092, Tacoma, WA 98416

Table of Contents

List of Figures

Acknowledgements

Thank you to my parents for finding ways to help me be curious. R. Lee Lyman was my graduate mentor, and he is an excellent scholar—one who embodies all of the effective habits that are described in this book. Thank you Lee for your patient mentoring. James Kennedy, Michael J. O'Brien, Michael Huston, and Barney Venables have also been important mentors, ones who helped me become a confident writer. Dana Lepofsky and Justin Nolan are two of my closest colleagues and friends, and their advice and mentoring have made me a better professor. Jon Dombrosky read and provided helpful comments on an early draft. I thank Marsha Quinlan for getting behind this book from initial submission to final publication; each step of the way her encouragement mattered and her advice helped get the most out of me as a writer. It is fair to say that this book would not have been finished without her support and guidance. My wife, Lisa Nagaoka, is a first rate scholar, and I am indebted to her for slipping ideas into my life with great kindness and stealth, such that I often think they are my own. I dedicate this book to all of the students whom I have had the pleasure to mentor.

Abstract

Entering a graduate research program in the sciences or social sciences—including the many fields that contribute to ethnobiology—requires a fundamental transition from being an undergraduate learner to becoming an independent, investigative thinker. The transition requires moving from structured learning using prompts for reading, writing, and exam-taking to an unstructured environment in which one is expected to know a field of study, identify significant research problems, acquire one or another form of data, and answer questions using skills and expertise to fill a gap in knowledge that is reported in a thesis or dissertation. An independent researcher can undertake the whole research process without prompts and has acquired the skills and knowledge to make contributions in a field of study. Faculty members and students often assume that gifted undergraduates have research skills, an assumption that students will be able to "figure it out when they enter graduate school." The premise of this book is that students do not have independent research skills at the beginning of graduate programs; indeed, the purpose of graduate education is to gain and hone those skills. Although there are many books about designing research, what makes *Becoming a Researcher* unique is its focus on this transition from structured learning to independent research. The book offers exercises on routine setting, time management, peer review, developing areas of research interest, and essay writing to focus on research topics and questions. Those exercises, moreover, are threaded within narrative about how to become a researcher, particularly concerning new habits and mindsets that must be adopted to succeed during the graduate career. Late in the book, lessons from the exercises are put into action to help students build products that lead to achieving milestones during the first year of their graduate program, particularly the design of a research proposal. At the end of the first year of graduate school, students should be well on their way to becoming a researcher.

Preface

I have mentored many graduate students during the last decade. Although my students come from diverse walks of life and range from fresh faces, just out of college, to experienced professionals with years in the workforce, they each confront a common struggle as they enter their graduate program—an intense transition from their previous learning or work environment to becoming a researcher (see also Davis and Parker 1997). I argue in the pages that follow that the most important transformation among successful graduate students represents leaning into the unstructured learning environment of research. Without exception, my non-traditional graduate students who come back to school after years in a profession had been working in a structured environment, one with set rules and expectations about work performance. A structured learning environment is even more prominent for undergraduate students, most of whom mastered test-taking, term paper writing, and class readings in a highly structured environment.

The work environment of research is nothing like the structured environments of undergraduate education and most professional workplaces. It is unstructured, which means that there is no syllabus provided by a professor or no work plan handed down from a manager. Faculty members take pride in the fact that they have accomplished independent research, and they expect that graduate students will engage in the same process if they are to obtain a graduate degree in a research program. The master's thesis or doctoral dissertation is the evidence that such a transition has been made. Through teaching the master's degree research-design course at my institution, I became aware that most faculty members simply expect students to "have what it takes" to do independent research once they arrive in graduate school. What I found, however, is that faculty expectations of incoming students are misplaced. Frankly, how could students be prepared to be independent researchers given their backgrounds in structured learning and work environments? Further, didn't they come to graduate school *to learn the skills of independent research?*

I have written this book assuming that all incoming graduate students are master's students; however, the lessons of the book also apply to incoming doctoral students, particularly if they did not earn a master's degree before beginning their doctoral program. I have thrown out the belief that even the best applicants to graduate programs are ready to do independent research; I have replaced it with a conviction that graduate education represents a transition from structured to unstructured learning. The student enters a graduate program not because she or he is an independent researcher but because that is what she or he wants to become. Thus, the expectation that a student will "come up with a topic and do research" is fundamentally flawed; they should be able to do this *when they leave graduate school,* not when they enter.

An important question is, "Why publish this book as a monograph through the Society of Ethnobiology's *Educational Contributions in Ethnobiology* series?" There are a few good reasons. First, the book grew out of teaching a graduate research design course in my home department, which is very interdisciplinary. The course reaches an audience of students with exceptionally diverse research interests, which forced me to think about key elements of the research process and to avoid the nuances of research in my own field. In my experience, ethnobiology requires a similarly broad perspective. Nabhan (2013:2), for example, refers to ethnobiology as an interdiscipline. Second, I was honored in 2016 with the Ethnobiology Mentor Award, and this book presents my strategy for mentoring graduate researchers. Third, in the world of academic publishing, there are many for-profit presses that benefit from scholarly articles and books. The Society of Ethnobiology is a non-profit organization that supports the *Journal of Ethnobiology, Ethnobiology Letters, Contributions in Ethnobiology,* and *Educational Contributions in Ethnobiology.* It is my hope that this book will attract new student members from across the various disciplines represented in ethnobiology and that income generated from the book will benefit the Society, particularly the *Contributions* series. Alternatively, if the book gains traction as a helpful introduction to graduate research, perhaps it will draw people into ethnobiology from other research communities. Finally, education in ethnobiology is increasingly important (e.g., Quave et al. 2015), and though this book is not precisely ethnobiological, it offers a foundation in research that I hope all new ethnobiologists will find useful. This book does not represent *the way* to do research, it represents *a way* to do so, one that I have found success with in my graduate research course and when mentoring graduate students.

In teaching research design, I noticed a few important habits in new graduate students. First, they routinely and enthusiastically name their field of interest when asked what they study. They almost never describe a gap in knowledge in the field and thus a research topic that they will focus upon. Second, they may show intense interest in a particular methodology, but if so, it is rare that they focus upon research questions that are of interest in their field using that methodology. Third, students are trained to always appear as if they are experts; thus, they may resist that they have much to learn. Indeed, the graduate application process may encourage this perspective because (as faculty members) we celebrate the arrival of new recruits; ones that we believe have high potential. A problem is what we evaluate in their application portfolios likely reflects intelligence and work ethic, but may not reveal preparation to do research.

An independent researcher is a person who knows a field of research very well, one who has read the literature and who knows current gaps in knowledge. Further, this person is able to articulate those gaps in knowledge as topics, can frame research questions and study objectives, and knows the appropriate methods to collect data. She or he also knows how to analyze those data and how to make interpretations that inform conclusions that fill the gap in knowledge. Most importantly, an independent researcher can plan the entire process from generating questions to data acquisition to report writing and presentation. That is, he or she can provide the structure for this unstructured process.

This book takes the student and professor through the transition from undergraduate learner to graduate researcher. It provides a narrative approach for diving into a field, zooming in toward a topic, framing precise research questions and study objectives, and developing a thesis proposal with a research plan. Along the way, the student will face milestones and will develop important relationships, particularly with their major professor, thesis committee members, and student colleagues. In addition to discussion of these challenges, each chapter presents exercises that enable practice and refinement of skills required to do independent research. The book focuses on the transition to graduate research during the first year of graduate school, up through presentation of the thesis proposal. Indeed, it is good news that a researcher only needs to go through this transition once; afterwards comes the enjoyable, satisfying life of being a researcher.

In my experience, graduate students underestimate the amount of effort that is required to become a researcher, particularly what is required in the first year of graduate school during which it is tempting to default back to habits that were effective throughout the undergraduate career. Because of this challenging transition, there tend to be four outcomes for students who enter a graduate research program:

1. They adopt many of the approaches recommended in this book and successfully make the transition from undergraduate student to independent researcher.
2. They squeak through with a weak thesis by wearing down their major professors and committee members, only to be unqualified for most research jobs.
3. They stay in graduate school much longer (e.g., many years) than they anticipated or than their major professor wants.
4. They become frustrated and quit prior to earning their degree.

Clearly, the first outcome on this list is what students and major professors want; what propels students toward the other, less attractive outcomes is that they underestimate what becoming a researcher requires. Establishing great habits early in one's graduate career is the key to success, but it often requires considerably more effort than students realize.

This book is intended for use during the first and second semesters of graduate school. It takes the incoming student through the thesis proposal defense at which point I assume that students have the ability and confidence to pursue the thesis research. The approach that I share in this book is the one that I adopt with all of my graduate students, most of whom have successfully made the transition to becoming a researcher.

Chapter 1: Becoming a Researcher

The incoming master's student is an interesting breed. There are at least three types of personal histories that lead one to seek a graduate degree: 1) an advanced undergraduate student may seek to continue to learn in their area of interest; 2) a career professional (analyst, educator, or practitioner in another area) may desire to build or enhance their skills; or 3) a person with no direction may seek to back into a graduate program because he or she is not sure what to do with his or her life. There are exceptions to these stereotypes, and incoming students may have a combination of motivations—at times feeling lost, at other times feeling curious and engaged. The current educational model in many university programs in the United States encourages high enrollments, and many master's programs draw incoming students from each of these (and other) histories. During the application process, the incoming student might not have been challenged to critically evaluate a variety of programs to determine which suits their needs. Alternatively, career professionals may not have the luxury of selecting among a high number of programs. Thus, there may have been little consideration during the application process about the purpose of a master's degree.

Master's programs are diverse. That said, it is rare that a single program can serve many goals. Some programs, for example, may focus on acquisition of skills, such as quantitative or qualitative methods, with little to no emphasis on research design and writing. A separate issue is that in the humanities, sciences, and social sciences universities may devote more resources to doctoral programs. Graduation of students holding PhDs often confers greater financial investment and academic reputation to a university than graduation of master's students. Thus, universities may not only invest more in doctoral programs, they might push for doctoral admissions over master's admissions. This would make sense if all students came prepared to do independent research or if all applicants desired to become researchers. There are numerous options for applicants who desire a professional degree that de-emphasizes research, such as a Master's in Business Administration or Education. But what about the student who desires to become a researcher? What should a master's degree look like for such a person? Should they skip the master's degree and target a doctoral program? Many professors feel that it is advantageous for students to move quickly into a doctoral program. I disagree because in my experience it is exceptionally rare that a student who desires to become an independent researcher is prepared to enter directly into a PhD program. In addition, many doctoral programs that admit students without a master's degree require the student to get one as part of their PhD degree plan. Regardless if one enters a stand-alone master's program or directly into a doctoral program, the fundamental transition is from undergraduate to graduate education.

There are a number of reasons that students push (or are pushed) to apply to doctoral programs without a master's degree. First, as mentioned above, universities may encourage it (even market for it). Second, there is little to no consideration by applicants of what it means to do independent research. Third, there may be little to no consideration by students or faculty members concerning the purpose of a master's degree. If a student is seeking an easy, accessible professional experience to build skills, then an online master's or one that requires little on the research front is sufficient. However, such a degree will not prepare most students for a doctoral program. On the other hand, if a student desires to make a transition from being an undergraduate learner (consumer) of information to an independent researcher who can solve problems through mounting evidence and designing logical arguments, then the right master's program can be essential preparation for becoming a career researcher (within or outside of doctoral research). This book is for those students. To get started, we must consider what independent research is.

To "research" literally means to "investigate again;" as in, to continually analyze material within a field of study. An independent researcher knows their field exceptionally well. She or he has read how the field developed, knows important theoretical premises held by members of the discipline, holds relevant analytical skills, and can write about their research in ways that others can understand it. Although each of these characteristics is important, what truly makes an independent researcher is even more basic: he or she can ask a question that needs answering in the field, has the skills to assemble data as evidence, and can construct a logical answer to that research question without much guidance from other members of the field. Most undergraduate students cannot do this. Thus, the master's or early-career doctoral student engages in a critical period of transition.

The transition is from what can be termed a structured classroom (or professional) environment in which learning tasks are outlined by someone else (e.g., a professor or manager) to an unstructured learning environment. Research is unstructured learning, and an independent researcher is able to envision *all* tasks that must be accomplished to answer a research question. Further, that person is fully capable of organizing the outcomes of research, writing and presenting about the results, and fully articulating the implications of a study for members of the field. The independent researcher does not require external prompts, schedules, ideas, or other forms of guidance to accomplish a study because they know how to assemble and structure the entire process on their own. A master's program that focuses on research should produce graduates who are able to produce independent research. I have yet to encounter a student with a bachelor's degree who is prepared for all aspects of independent research. A good master's program can provide this transition. Unlike many experiences, "there is a there, there" when one seeks to become an independent researcher. So, how does one get there?

The first step is one of becoming aware. An undergraduate education is student centered, but research is about the research product, which for a research master's program is a thesis. A master's thesis is a body of research that addresses a topic within a field, asks one or a few important questions that need answering in that field, and through providing those answers solves a research problem by communicating new understanding—that is, the research fills a knowledge gap. The incoming student starts from the much-encouraged social belief that education is student-centered, which represents an undergraduate, structured learning environment that is comparatively comfortable. She or he will immediately confront a new, faculty-held belief system that learning is idea-centered, which represents the unstructured learning environment discussed above. Thus, there is a transition in awareness about what education is about; becoming an independent researcher is a different educational goal than choosing an undergraduate major and learning fundamental information in a field of study. The sooner a student confronts this transition, the better. Indeed, a lack of such awareness leads to an onerous master's program with delayed or perhaps no graduation, a process in which a student will likely come to resent the research process rather than embrace and enjoy it. Thus, step one is to establish awareness that a change in values about education is required to make the transition from undergraduate learner (or managed employee if one is coming back to school from the work force) to graduate researcher.

Step two requires additional awareness that the transition is at times uncomfortable, which can be countered by establishing routines. During the graduate career, successful students will establish multiple routines including those centered on learning skills through courses, reading in one's field, and writing. Graduate courses offer somewhat structured learning environments that provide basic concepts and skills, but course selection requires that the student and their major professor thoughtfully explore and design a degree plan. Most master's programs have established policies that guide students toward a major professor, so I will focus on routines that are much harder to establish—reading and writing. Indeed, routines of writing and reading are commonly ignored in master's programs, and it is often assumed by faculty members that good students will set their own goals, prioritize reading and writing tasks, and produce a research proposal followed by a thesis. Since incoming students have almost never engaged in this process, which is the bread and butter of independent research, it is an awful lot to assume that they will "just learn it along the way." This book is dedicated to setting reading and writing routines through a series of exercises that help student focus their interests because the assumption that a student should (or even can) learn those routines on their own is flawed.

A third important step is to create a support network. Setting a routine and making the transition to becoming a researcher is an uncomfortable one for many students, and because students enter graduate programs in cohorts, whether or not one emphasizes a supportive or negative social environment matters. When students enter a program as a cohort, they share a similar level of experience; they may also share office spaces, classes, group projects, meetings, and a number of other experiences. From the beginning, whether or not one seeks to create a supportive environment for setting routines, prioritizing reading and writing time, and making the transition to an unstructured learning process is very important. It is inevitable that entering graduate students will establish a new social network. This can either be negative and based on shared complaints about the challenges of becoming a researcher, the workload, and the program standards, or it can be positive and encouraging. The latter type of network is based on discussing shared challenges and providing solutions. From a faculty member's perspective, it is entirely up to the student as to which type of social network she or he establishes.

There are a number of other challenges that new graduate students will face, and these are covered in the following chapters. The remainder of this chapter is dedicated to setting up a reading and writing routine, which is fundamental to success and should be developed right upon beginning a graduate program. This book is organized around a set of exercises; this first section centers on starting and developing an annotated bibliography, which is a reading and writing process. Chapter 2 covers time management and centers on time awareness, short-term planning, and long-range planning exercises. Chapter 3 orients the student toward the fundamental goal of this book, which is to lead him or her into and through the process of developing a thesis proposal. To develop a proposal, the student must engage writing about their field of study in general, gradually narrowing to multiple potential topics of interest. Gaining focus and choosing a topic for research requires an informed commitment, one that can only be made securely once a student gains fundamental knowledge about their field—which is the purpose of starting with the annotated bibliography exercise. Along the way, students will need to establish a support network, which includes establishing a cohort and a thesis committee; I cover these topics in Chapter 4. The commitment that was required to select a topic is refined and renewed with the selection of one or a few important research questions, the answers to which fill knowledge gaps or solve practical problems in the field, which is the focus of Chapter 5. Once the researcher has a clear topic with related research questions, it is much easier to develop project objectives and a research plan. In addition, the student is prepared to construct a centerpiece of their thesis proposal, the literature review. The literature review is the subject of Chapter 6, and the transition from research plan to thesis proposal is the focus of Chapter 7—an important exercise at that point in the process is to construct a proposal storyboard. The written proposal, however, is only part of the requirement; the student must give a presentation about their proposal, which is the subject of Chapter 8. The book ends with the thesis proposal; the thesis, however, I leave to the student and the major professor (a.k.a., advisor or thesis chair). Along the way, I present sections of chapters that cover small hurdles, such as dealing with frustration or communicating with peers and mentors. Thus, this is not a proposal- or thesis-writing book; it is a book about becoming a researcher. During that transition, a student should be increasingly prepared to accomplish proposal and thesis writing.

At the end of the book, there are three appendices. The first includes two types of preproposal exercises, those that are presented in the chapters 1 through 6 with more-detailed instructions and a few supplementary exercises. The second appendix is a copy of the graduate student handbook for the master's program in my department; the purpose of presenting the handbook is to highlight that the challenges discussed throughout the book are designed into the structure of a graduate program. The third appendix presents exercises that relate directly to proposal writing relating to chapters 7 and 8.

I have drawn heavily on a number of sources during the writing of this book, each of which are companion volumes. Precise distinctions between research topics, questions, and problems are available in Booth et al.'s book, *The Craft of Research*. Routine setting and other advice on writing draws from *How to Write a Lot* by Paul Silvia. *Writing the Doctoral Dissertation* by Davis and Parker was an important guiding resource during my graduate career and offers sound advice on time management, thesis committee structure and roles, and proposal writing. It is fair to state that I would not have written this book without having read those books and many other resources, which are listed in the "important sources" section at the end of this book. The reader should consider each of those sources at the beginning of their graduate career; I use those books as reading material in the graduate research design course that I teach to entering master's students each fall.

Starting a Reading and Writing Routine

In his book *How to Write a Lot*, Paul Silvia makes a concrete argument that the single most important practice for a researcher is to establish a writing routine, to put it above all other commitments, and to stay dedicated to it. The fundamental reason this routine is critically important is that research writing is very difficult. Silvia illustrates that research writing is direct, information-based, and at times tedious and boring. He falsifies a number of "specious barriers" to writing, such as the need to "find" time and the need to feel inspired. Writers who wait for time to become available and for the "muse to speak" become binge writers—those who cram writing into short, unreliable blocks of time—at best. More likely, however, is that those who give into specious barriers may never write at all. In establishing a support network, the graduate researcher should pay special attention to the need for support in setting up and sticking to a writing routine. At the onset, this is tough to do, but the advantages are significant.

Writing is a cognitive, creative exercise that is done in one's own voice. If a person cannot articulate ideas, concepts, and arguments in writing, then he or she does not yet know the subject matter. Thus, *writing is practice with information* from the field within which one is becoming an expert. Routine writing is not something to wait on; it represents a cognitive process that should be part of the entire research career. Additionally, when a researcher writes on a routine basis, the products build up over time. Even writing summaries of articles that one has read comes back to aid in constructing the proposal and thesis. When you write about someone else's work, you do so in your own words; this process of articulating someone else's research in your own voice increases your understanding of your field and enhances your ability to identify gaps in knowledge. Routine writing—particularly if one keeps a daily record of it—is satisfying. The milestones of accomplishment in graduate school are few with long periods between them, such as defending the proposal, defending the thesis, and graduation. If one does not set short-term goals, the process can become discouraging.

When I first read Silvia's book, I was within a period in which I was choosing not to write about research. I recognized right away that the most productive periods of my career for proposing and receiving grant funding and writing articles and books were ones in which I followed precisely the routine that he describes. He recommends that a researcher set aside a regular block of time at the same time each day for a committed number of days each week. I currently adopt a routine of writing from 7:30 to 8:30 am each morning. Before receiving tenure, I wrote from 7 to 9 am each day with my office door closed. It seems harsh, but my thoughts were, "students can have crises after 9 am, when my office door opens." For me, the rest of the world does not exist during my writing time; if someone calls, I do not answer. I do not check email, and I send students away. Writing early in the day reduces the opportunity for distractions and is a period during which I have energy and creativity. It is important to gauge which part of the day is the best for writing for you and to prioritize setting that time for your routine writing.

If you are imagining that becoming a graduate researcher is a serious commitment, you are correct. However, contrary to common practice among students (and many faculty members), what I just described is the easy way. Without a routine, you should not expect to write very much. If you need a clearer message than that, read Silvia's book.

If you are a beginning researcher, you may be wondering, how can I write if I have nothing to write about?! This too is a specious barrier to writing. There is always something to write about, but here's a real boost—start an annotated bibliography right now!

The Annotated Bibliography

The most significant aspect of the transition from structured problem solving (either in education or in employment) to unstructured learning through research concerns becoming an avid reader of the literature in your field of research. Equally important is becoming a research writer. The two are intimately related, as reading from scholarly journals is an encounter with the research writing of others. Research writing is difficult because it needs to be factual and concise; thus, it is by no means poetic. This is not to say that research writing must be boring; as one becomes a better research writer, reading scholarly publications becomes increasingly interesting. One begins to appreciate papers, chapters, books, and other examples of scholarly writing that are well crafted. Indeed, research papers (and other forms of writing) are *designed*. As a beginning graduate researcher, it is important to dive into the literature in your field right away. A big part of this challenge is knowing where to start as your research interests are likely to be unfocused. Recall, however, that establishing a routine is a critical component to becoming a research writer, so a productive exercise is to set goals about how many articles or chapters you will read each week and to write summaries of them.

An annotated bibliography is a collection of those summaries (see also Appendix 1, Exercises 1 and 2). It should include:

1. a full bibliographic reference for the source,
2. a summary of what the research is about,
3. a description of the research questions that are addressed,
4. a characterization of the main claims that the authors make and the evidence on which those claims are based,
5. as well as a description of how the source fits within one's research interests.

The same format should be used for each annotation; digitally archive notes and pdfs of articles and chapters together.

In the process of reading and writing for the annotated bibliography, students often make several mistakes. First, many students do not know how to give sources a cursory read, which is deeper than skimming a source but shallower than a thorough reading. The point of reading for the annotated bibliography is to determine the basic meaning of the source related to one's own interests. The researcher may even want to make the last component of the annotation be a reflection on whether or not the source requires a deeper read, which can then be planned for a later date. A second mistake is to conceive of the annotated bibliography as work (or worse, as an assignment); instead conceptualize the annotations as what they are meant to be, *your personal notes about the source.* The audience of the annotated bibliography is none other than you, at a later date. Precise, informative notes about a source that provide reflection on how the research relates to one's own interests are critically important for future use. One does not want to confront vague notes, inaccurate citation information, and disorganized thoughts in the future. Students can easily incorporate reflexive notes in digital citation programs, such as Endnote or RefWorks; the point is to *start the routine* of reading and writing. Here are a few strategies that one of my former graduate students uses for annotations:

1. Read over the introduction to a chapter or articles and identify the major questions addressed in the work.
2. Read the methods and try to figure out how the authors answered the research questions she or he asked.
3. Understand the figures and tables presented in the work. Many times large sections of narrative are based on charts, maps, illustrations, and summaries of data.
4. Read the discussion and conclusion carefully. Think back to the introduction and methods; did the author accomplish what they set out to do?

Most entering graduate students aspire to become a researcher, but they typically have little to no research experience and have not read the literature. Starting a routine of reading and writing for the annotated bibliography "kills two birds with one stone." The routine should include a few important components: 1) the student should choose and schedule times of days and days of each week that will be dedicated to reading and writing; 2) a reasonable way to set a weekly goal is to choose the number of sources that will be annotated each week; and 3) the student must set expectations that are practical, such that weekly goals can be met. The student must be devoted to this routine, must schedule it such that it does not compete with other life responsibilities, and must establish a pace that can be accomplished. The goal of routine-setting is not to produce super-human effort each week; the goal is to spread the effort of reading and writing out and to make the practice of each habitual. It may take a few weeks of trial and error to establish a balanced routine, but if balance is not the goal either too little will be accomplished or expectations will be too high. Both of those lead to a discouraging experience; in contrast, accomplishing a balance provides achievement of incremental milestones, which is satisfying. Without such shorter-term milestones, research accomplishments, such as finishing proposals, theses, and articles, require long periods of investment between accomplishments. Thus, the reading-writing routine of the annotated bibliography is a productive, healthy research practice that will pay future dividends when major milestones are reached.

Even when a routine is established, other challenges will confront the beginning researcher who is likely to encounter a quagmire of articles, chapters, books, dissertations, and other resources in their field. This is intimidating because one likely does not have the research focus to choose which resources are significant. One of my students aptly described this challenge as "sailing a boat while one is still building it." There are several ways to confront this challenge. First, be aware that in surveying the literature in one's field, beginning researchers will encounter sources that will not be used—metaphorically, "one must kiss a lot of frogs to encounter the prince (or princess)." Students can expect this to be a frustrating part of the process; however, some of those sources may be of future importance in ways that are difficult to envision, so it is important to annotate even those sources for which one does not see an immediate use. Second, once a beginning researcher starts the process of creating an annotated bibliography, she or he is essentially beginning the process of a literature review, which will be important for the proposal and the thesis. During annotation of those sources that students recognize as important, gaps in knowledge presented by authors should be identified. These might serve as ideas for focusing one's research topic as the thesis proposal is developed. Third, related to those sources that are of interest, the student should establish a *bibliographic trail*. What other sources are cited in articles, chapters, books, and theses that should be read? Which scholarly journals publish commonly cited articles? What key words are used to describe the sources that one finds interesting? Following that trail will increase one's focus. I tell my students that they should develop their own list of key words for digital searches for sources. Five to ten appropriate search terms can significantly narrow the resources one discovers.

Finally, it is critical that the graduate researcher realize that they are not alone. The most important resource available to each student is his or her major professor. When beginning the annotated bibliography, a great starting point is to ask one's major professor what are the top five sources I should be reading this week? In the following week ask for five or ten more; then, start that list of your own key words, begin a bibliographic trail, and most importantly set a reading and writing routine—the most important component of your transition from a structured education to independent research.

Summary

In this chapter, I have made a few critical points about becoming an independent researcher as a beginning graduate student. If a student is an incoming doctoral student, she or he may face many of the same challenges as an entering master's student. Graduate programs are diverse, and students may or may not have considered whether their career aspirations match a particular program. Some graduate programs do not hold research as a focus. In research graduate programs, faculty members hold particular values about independent research—mainly that it is an unstructured learning exercise. Many faculty members (accurately) lament that incoming graduate students are not capable of directing their own research, which indicates that there is a gap between the expectations of the major professor and the student. I argue that the master's degree (or the early years of a doctoral degree) represents an uncomfortable transition from being an undergraduate learner in a structured environment with prompted assignments, exams, and other exercises to becoming self-directed within an unstructured research environment. Graduate students must be aware of this transition and lean into it if they are to become independent researchers capable of filling a knowledge gap in their field. Third, reading and writing about the subject matter of one's field is a critical component of being a self-directed researcher. Students may find it difficult to engage the literature as he or she may not yet have a research focus. The student, however, must start somewhere. A great starting point is to ask one's major professor for a list of critical sources to read and to begin assembling a list of key words that relate to research interests. To begin a writing routine, piggy back an annotated bibliography onto the readings. Make this a habit, and by the time you have developed a topic and research question you will be well on your way to becoming self-directed.

Establishing this routine is only the beginning; the student will need to triangulate their research toward a topic within their field and to articulate research questions and objectives to write their thesis proposal. This entails that the writing and reading routine grow and change, which is the subject of later chapters in the book. To be successful, the student will need to adopt a time management approach that identifies tasks, establishes priorities, envisions time commitment, and allocates time effectively, which is the subject of the next chapter.

Chapter 2: Time

Establishing a reading and writing routine is critical for success at becoming a researcher. That routine must fit within a bigger picture of time management. The new researcher must overcome the specious barrier that Paul Silvia writes about in *How to Write a Lot* that it is difficult to "find enough time" to read and write. If one conceives of time as a resource that has to be found, one will not set and prioritize a routine. Someone who consistently confronts this barrier actually may not want to write because it is difficult, a hidden fact that underlies their time attitude. Britton and Tesser (1991) discuss three fundamental components of time management—time attitudes, short range planning, and long range planning—each of which influence a student's willingness to invest effort in their work. A time attitude is one's demeanor or feeling driving his or her desire to spend time addressing a challenge, such as reading sources or writing about research. Giving into a specious barrier, such as the belief that one cannot find enough time, supports a negative time attitude, one that leads to a quagmire of low reading and writing productivity. For our purposes, it is simply important to recognize that one has a time attitude, that choices influence it, and that one can change it.

Britton and Tesser's study focused on undergraduate students and concerned their study habits; the authors found that time attitudes were strongly influenced by short range planning, which encompasses delineating a series of tasks to focus upon for a week and within each day. When students broke down their study time into short timeframes, such as daily chunks aggregated within study weeks, they were able to maintain a sense of accomplishment from gradual and sustained completion of schoolwork. Establishing a daily reading and writing routine under the umbrella of a workweek does the same thing for the researcher. Indeed, should a graduate student set up such a routine for himself or herself, he or she has brought structure to one component of the unstructured learning environment of independent research. The ability to accomplish tasks has a positive influence on one's time attitude, making the challenge of reading and writing less intimidating and eventually making it a rewarding, satisfying experience—one that reinforces itself. Thus, establishing meaningful tasks and sub-tasks within one's routine is critical, which is returned to in the next section.

Long range planning was not as impactful on undergraduate time attitudes. One might plan how to complete the semester, anticipate personal rewards for good grades, or establish a timeline for graduation, but these have little effect on the student's attitude toward her or his daily and weekly routine. Satisfaction related to long range planning is decoupled from the time investment in study habits that are required to get there. Thus, it is the short range planning that has the biggest influence on one's routine. The same can be said for establishing a reading and writing routine as a graduate student. Filing one's degree plan, establishing a thesis committee, defending a thesis proposal and eventually a thesis, which are followed by graduation, take months and years to accomplish; one's daily time attitude is simply not affected by these experiences. Without short range planning, all that a researcher has is the hope that someday their proposal or thesis will be done; it is no wonder, under such circumstances, that a person feels they must "find more time." Ironically, all the student researcher has is time; what is missing are the habits to support a time attitude enabling accomplishment.

It is important to recognize that developing a healthy time attitude does not come from working all of the time. For years in my statistical research class, I have referred to two kinds of fun. For students, there is fun that is guilt free and then there is dark fun. If a person has a stable routine in which he or she makes daily progress toward short-term goals that fit within a longer-range plan, then it is easy to step away from graduate research for an evening or weekend of fun. However, if all that a student experiences is fear and stress in the absence of a routine and the accomplishments it supports, then it is difficult to step away because very little gets done. Indeed, students convince themselves that "they need a break," or that "once they return they'll finally make some progress." The need for a break may become constant work avoidance. In the following weeks and months that person will find that they continually "need a vacation to escape the stress of graduate school" because without short range planning and a shift in time attitude resulting from establishing a routine, nothing will change and very little is accomplished.

Finally, here is a note on long range planning, which is important to mention before discussing how to delineate and prioritize tasks and sub-tasks. It is important to "keep one's eye on the prize." I mentor several undergraduate and graduate researchers, and each semester we write semester goals that refer to degree milestones on the grease board in our lab (Figure 1). However, we only refer to those goals three times during the semester: at the beginning when we envision those goals, at mid-semester to deem if they need adjustment, and at the end to determine whether or not they were accomplished. Yet, the long-range goals are there for the entire world to see, and each week during our meetings, each of us delineates our daily and weekly plans, which we review the following week. We discuss our progress publicly, which destigmatizes a bad week and provides social support for good ones. The result is that our short range planning is embedded within long range planning, and our goals are accomplished.

Time Awareness Exercise

People treat time in much the same manner that they treat food; thus, it is difficult to change one's eating habits or one's time attitude. As with food, people are often unaware of how they manage their relationship with time. I recall an episode of Jamie Oliver's Food Revolution, a television show in which the British chef desired to help US schools counteract obesity in children through changing school lunches. In one episode, Oliver arrived to a school, did an assessment of lunches, taught school chefs to cook healthier options, and presented

Figure 1. The grease board in our graduate lab where we post weekly (left) and semester (right) goals. Students affectionately nicknamed it the BOS (Board of Shame).

students with healthier foods. He found that there was little buy-in and that resistance to change was systemic. Students, chefs, and even parents did not like the stress that a change in diet produced. Oliver was passionately concerned about the poor quality foods that students gravitated toward; in frustration, he went to a dramatic length to show them the extent of the problem with the school's food. Coming back from a commercial break, the viewer (and the parents and teachers) were confronted by a dump truck backing into an area of the schoolyard. It dumped a full load of meat fat in front of the community members; Oliver emphatically stated that this was the amount of fat served in school lunches each week. Made aware, adults in the community embraced change, and students began to follow suit.

Student researchers should do a similar exercise with their time (see Appendix 1, Exercise 3). For one week, document how each hour of time is spent. Be detailed, and be honest. Create a spreadsheet showing where each hour goes. Is there time spent out at a bar? Record it. Watching TV, eating out, walking to school, going to the gym? Record them. In addition, reflect upon one's long-range plan for graduate school; write a daily journal about what one hopes to accomplish. At the end of the week a student will see how their time is allocated; as that person has also recently journaled about their long term goals, they will easily be able to see whether or not much (or any) of their time was spent on tasks related to those long term goals. Finally, assess how time can be reallocated to create the balance of establishing reading and writing time. Once this awareness has been gained, a graduate researcher should be ready to determine tasks and sub-tasks as part of short range planning so that she/he can establish her/his reading and writing routine.

Task Delineation

I have emphasized the importance of establishing a routine for reading and writing early in this book because the milestones of graduate thesis research are few and far between. A critical component of establishing a successful routine is envisioning tasks, an important aspect of short range planning, which is a critical driver of one's time attitude. In my experience, when students first set weekly goals, they set aggressive benchmarks with too many tasks that are too large. This represents a "more is better" approach, but it is fundamentally flawed because delineating too many tasks that are too large leads to routine disappointment. On the other hand, students may adopt a low-ball approach in which they delineate too few tasks that are too minor. What should a list of tasks look like?

First, task delineation must be strategic. Any task must have a clear relationship to bigger goals and fit into a long-range plan. Second, any task must be something that can be accomplished within an allotted period. Third, tasks should be set weekly and daily. Make a commitment for what workdays will incorporate one week at a time. I meet with my graduate students once per week (at a routine time, I might add), and we list the tasks we wish to accomplish. We criticize each other's lists. A task that is too large and vague, for example, is "draft my proposal." There are many tasks that go into proposal writing, such as reading sources, composing a literature review, conceptualizing research objectives, or learning a methodological approach, thus setting the proposal as a single task is impractical. When students set a task that is this large, it will stay on our grease board under their weekly priorities week after week and perhaps for months. When asked, "How is it going?" The student is likely to respond, "Well, I didn't make as much progress as I hoped, but I worked on it this week." The task becomes a quagmire of disappointment, and eventually the student's time attitude declines to the point of discouragement and she or he avoids the task. I have witnessed this countless times. Thus, when a student lists such a task, I immediately state that it is "too big."

It is tempting to lean too far in the other direction using the low-ball approach to task delineation. For example, a student might prioritize, "write one paragraph of my literature review" or "read two articles" for a week, neither of which requires much effort. Alternatively, students might list tasks related to their coursework. Classes represent a structured learning environment; we limit task delineation to unstructured learning of thesis research during our weekly meetings.

The most effective way to delineate tasks is to connect them to the goal of *having a reading and writing routine*, which is directly linked to the long term goals of knowing one's field, constructing a literature review, determining a knowledge gap, proposing thesis research to fill that gap, and pursuing research toward well-defined objectives. Students who adopt this approach frame weekly tasks that make routine progress, such as "write for one and a half hours for five days of the week," and "read two sources per day for four days a week." Others set word count goals for writing periods. My current writing routine is for one hour each morning during the workweek. As Paul Silvia points out, such a routine must be "protected time" that is placed above all else in one's daily schedule. When "establishing the routine" is the commitment, productivity in reading the literature and writing gradually piles up. It is surprising how much can be accomplished in just one week with such a routine.

For example, I made a commitment to write a chapter on my field that was due in December of 2016 thinking it was to be a short summary of just a few thousand words. My plans had to change when I learned that the editors envisioned a chapter between ten and twelve thousand words. Caught by surprise, which was due to my own oversight, I did not prioritize the chapter into my weekly routine until the middle of October. I set a weekly routine, however, of one and half hours of work five days per week, and I finished the chapter on time, including editing, revision, and proof reading. The prospect of finishing the chapter was daunting until I committed to the routine; I was uninspired to write the chapter and wanted to focus on other tasks. The process of having a routine brought me into the chapter, which led me to become more excited about writing it. It was satisfying each week to review productivity.

Two additional parts of task delineation are important when establishing a routine. When possible, envision sub-tasks. I knew, for example, that I wanted the current chapter to focus on the topic of time management. I also knew that I would commit to a writing routine. The chapter is not monolithic, however, as it has multiple parts. Thus, I determined what the sections of the chapter would be; each day I chose which section to work on. Further, each section has goals related to communicating particular subject matter. In a sense, paragraph construction related to section goals became sub-tasks within the larger task of composing a draft of the chapter. Similarly, editing for clarity and flow is another step in the process, one to which I dedicate whole writing periods. At times, it is clear that a chapter is poorly constructed and must be revised; I focus on what bothers me about the current version by asking questions like "what is unclear?" Then I spend a period revising a section or multiple sections, which might include reordering paragraphs, deleting poorly constructed text and rewriting it, or adding transition sentences between sections.

It is also important to go beyond setting tasks by monitoring one's routine, additional sound advice from Silvia's book *How to Write a Lot*. Monitoring is a critical component of establishing a routine; my graduate students do this to some degree each week by discussing whether they accomplished their tasks delineated during the previous week's meeting. Ideally, this is accomplished by recording progress from each period; otherwise, it is more difficult to see progress. Monitoring of writing can be done in a number of ways, such as recording section completion. Alternatively, one can use the word count function on their word processor to keep track of how much she or he drafted in a writing period. For reading, monitor the number of sources as well as their annotation per reading period. Each morning, I construct a short list of tasks and sub-tasks for my writing period and cross them off the list when I finish them. I had not used this approach before writing the chapter on my field that I discussed earlier in this section. Monitoring enabled a sense of satisfaction each day as I recorded the numbers of words I wrote in drafts, the time I spent revising, and the sub-tasks and tasks that I crossed off a daily list.

At this point, you may be thinking something like "wow, this seems like an awful lot of attention just to establish a routine!" You are correct; it is a lot of attention to routine setting (Appendix 1). It gets down to whether or not one wants to become an independent researcher who knows his or her field well, can construct a research argument, can write to communicate with their field members as an audience, and complete a thesis in a timely manner. In a world in which many students languish in graduate school, this approach works—that is, it is well worth attention. More importantly, however, is that once established having a routine makes research far more enjoyable. There are additional ways to succeed with time management, which include prioritization, establishing stopping rules, and learning to do rest work.

Prioritization

During graduate school, students typically write research proposals, read research articles, present at academic conferences, and perhaps assist in teaching college courses for the first time. In fact, the incoming student might confront all of these challenges during the first week of their first semester. To strike a balance, responsibilities require management. Classes and teaching assistant duties are structured, such that a student should be able to examine their schedule and know precisely when blocks of time are unavailable. However, what tasks should you prioritize once you recognize visible blocks of time? When a student sets up a list of tasks each week, it represents a commitment. Those tasks that the student must do will be on the list along with estimates of time allocation.

Priorities will change during graduate school; early, the student is trying to determine what topic they should focus upon for their thesis research. Thus, graduate students starting on a topic should prioritize reading the scholarly literature in the field and writing summaries that make up the annotated bibliography. A reasonable routine for reading and writing is critically important because it dilutes the intensity of the prioritization process over many weeks and months. A student will confront periods during the semester in which they need to prepare for class exams, write term papers, and tackle a burst of teaching or research assistantship duties. If the student has a routine established, she or he is slowly piling up reading from the literature, which leads to gaining experience in the field. The student is also accumulating writing experience through annotating sources. If one knows that he or she will be reading and annotating five or six articles or book chapters in a week, it becomes reasonable—as one's experience grows—to understand the time investment required each week for those tasks. In addition, should a week come along in which other responsibilities are overwhelming, the student can reduce the number of sources to three or four. The reduction is reasonable because the student has gained confidence that during an average week she or he will read and write more. Thus, early in the graduate career the reading and writing routine should prioritize enmeshing oneself in the field's literature at an even, consistent rate.

As a student progresses through the first and second semesters, he or she can expect an increasing need to write. The following chapters cover strategies for triangulating from general interests to targeted research questions and objectives, but as one becomes more focused, writing becomes increasingly important. Thus, contributing to the annotated bibliography will eventually slow down. The student may add one to three sources to their annotated bibliography each week and shift toward writing a literature review, crafting narrative that explains the gap in knowledge to address in his or her own research, illustrating important methods to use and defending why their use is appropriate, and other components of a thesis proposal. The punchline of prioritization is that the student must gauge each week at what point she or he is in the process of thesis research. If one has not yet identified a research topic, then more time should be spent reading and annotating source literature. Once one has articulated a topic and research questions, one starts to shift toward drafting a proposal; then more time will be spent writing. The key is to set the routine; otherwise, the tasks of writing and reading will be deprioritized and forgotten amongst the more immediate responsibilities of coursework and teaching, not to mention the need to balance in time for friends, family, exercise, rest, and other important aspects of a healthy life. One challenge of such a routine, however, is that even best laid plans are at times tough to follow. One essential tool for success is to learn the value of rest work.

Rest Work

In their book, *Writing the Doctoral Dissertation*, Davis and Parker cover the details of the types of challenges graduate students will face in doctoral programs from routine setting to how to establish the doctoral committee and what the role of committee members should be to many other aspects of dissertation research. One exceptionally helpful concept that they describe is "rest work," which consists of tasks that relate to completing drafts of research documents, but the tasks are not actually composition of narrative (Davis and Parker 1979:31). All writers face that there are days when composition is simply difficult, but that does not mean that writing progress should not be made during one's scheduled writing period. There are numerous non-writing tasks that are important for completing articles, book chapters, thesis sections, proposals, and other documents.

All research writing requires the assembly of a bibliography. Bibliographic management software, such as RefWorks and Endnote, make this process easier than in the past, but all references should be proof read for small errors. A pesky challenge confronting many researchers is that each scholarly journal tends to have its own nuanced style guide; thus, in addition to the bibliography, writers must format headings, fonts, punctuation, and other aspects of style to finalize submissions. Universities require particular styles for completion of theses and dissertations; editing for style requires very little creativity and is work that must be done at some point. Similarly, production of maps, illustrations, charts, and data tables all qualify as rest work. Crafting charts, for example, may require diligent effort but a different type of creativity than composing narrative. In some cases, producing maps, charts, and illustrations may inspire creativity for composition.

One type of rest work that is critical for progressing through drafts is editing, which consists of reading sentence by sentence to determine if what has been written makes sense. Revision is more extensive than editing and usually takes place in response to comments provided by reviewers, the major professor, committee members, or other readers who have evaluated one's writing. Revision may require composition of new text, re-ordering or re-structuring of sections and paragraphs, or deletion of portions of narrative, which means that it requires creative energy. Nonetheless, editing and revision are not the same as composing first drafts and thus represent different processes that qualify as rest work. The point of rest work is to take advantage of writing time when composition is difficult and creativity is low. In contrast to rest work, stopping rules confront a distinctive issue of determining when to stop reading and writing within or at the end of a project.

Stopping Rules and Windfall Writing

I conceive of stopping rules in two ways. Typically, stopping rules provide an endpoint for a chapter, articles, thesis, or other pieces of scholarly writing (Davis and Parker 1997:31–32). A second way to conceive of such rules is to determine how much work is enough for a writing period. The first type of stopping rule relates to long range planning, and the second relates to short term planning. For long range planning, it is important to have a sense of what the boundaries are for a chapter, paper, or section thereof. As writing takes place, researchers commonly think of new, related ideas or obtain new data; in excitement, it is easy to think that those ideas or data should make it into the working draft. This may not be the case, but it is easy to determine whether to incorporate new material into a working draft. First, the writer must ask does the new material fit within the scope of the paper; that is, does it enhance communication of the main point of the work? Even great ideas that relate to a topic might not belong in the current working draft if they do not enhance communication. Second, and particularly for new data, the writer must ask, does incorporation of the data change the results? If the answer is no, then perhaps it is not necessary to incorporate such data once drafting has begun. Alternatively, if it is easy to add new data, then perhaps it is worth the time. If new data change the outcome of the study, then include them. The point is, the author must explicitly consider whether or not new ideas and data are worth the delay in completion of writing, particularly if new material detracts from the work's objectives (e.g., is distracting or tangential) or if nothing new is added.

It is important to jot down notes about a new idea for future work and then to return one's focus to the matter at hand—writing during one's allotted time. It is important to know when to stop adding new material; stopping rules should be guided by the study's objectives.

It is also important to develop stopping rules for short term planning. How much writing is enough for a given day? A routine only works well if a researcher sticks to it. If one is consistently writing outside their scheduled time or writing beyond it, the risk is that he or she will become fatigued and lose motivation. Most days when I reach the end of my writing period, I finish the paragraph or thought that I am composing, leave a note as to what should come next, and stop writing. I do this so that I can come to depend on the routine for helping me toward finishing drafts. That is, in the short term, I stop at the allotted time to come to trust that having a routine will lead to greater productivity than would writing in large, effusive blocks. When one only writes intermittently during large windows of time, it becomes binge writing, which is the opposite of a routine. Short-term stopping rules reinforce one's reading and writing routine.

Periodically a researcher will have the time and the inclination to write outside of the writing period. Paul Silvia refers to this type of writing as windfall writing; it represents extra productivity that should be treated as a bonus. If one is dedicated to their routine and keeps it, then windfall writing is a pleasure. Do not, however, get into the habit of replacing routine writing with windfall writing, which will erode the reading and writing routine.

Monitoring and Self-rewards

An important component of establishing a reading and writing routine is to keep track of progress (Appendix 1). This may seem like an unnecessarily detailed addition beyond staying dedicated to a routine; however, being able to visualize one's progress is a great way to stay motivated. I added monitoring to my writing routine at the beginning of writing this book after reading Silvia's book, a step that I had not taken previously during my research career. I keep track of my writing word count during my writing time at the beginning of each workday on a small grease board in my office (Figure 2). There are two aspects of monitoring that I find motivational for continued writing. The first is that when I do not feel like writing on a morning, I do not like the thought of a blank space on the board for one day that week. It is remarkable that there are days on which I am unmotivated, but just the commitment to sit down and write something—to get a word count on the board, no matter how small—turns the tide and leads to a productive writing period. I find that whether or not I want to write does not relate to whether or not I will be able to have a productive session once I get started. Sometimes it is tough to get started and monitoring helps clear that hurdle.

Monitoring is motivating in a second way; at the end of a week, and particularly at the end of a month, it is exceptionally satisfying to be able to look at the cumulative series of word

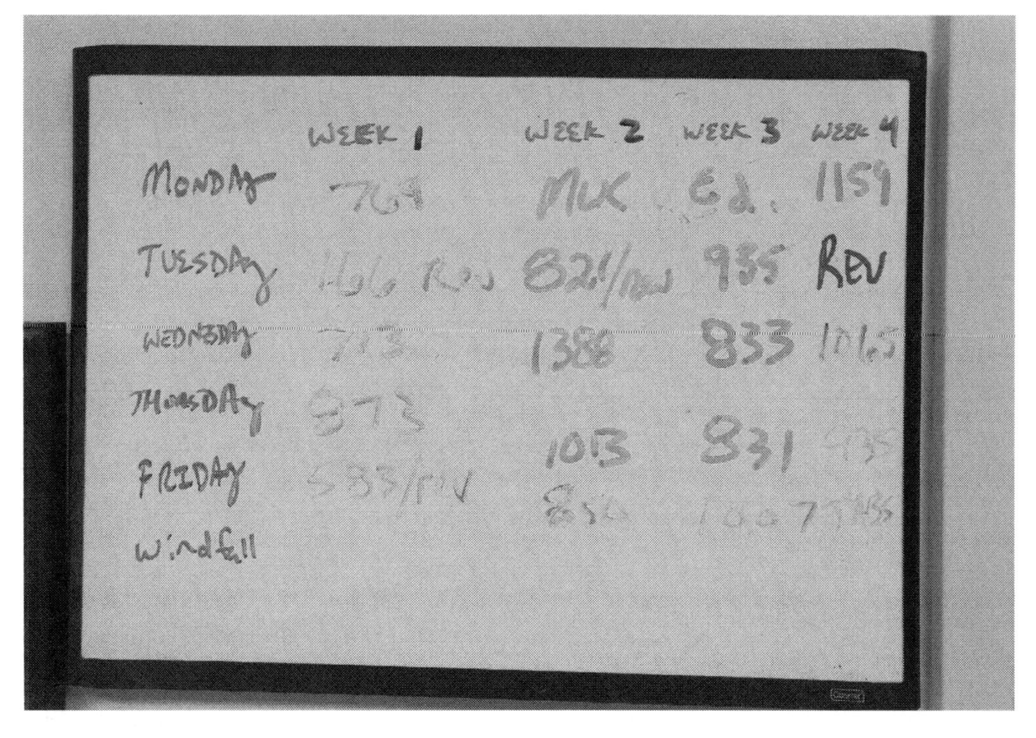

Figure 2. My word count monitoring board for one month during the spring semester for corresponding morning writing periods.

counts. The truth is that routine writing adds up, and one steadily makes progress on chapters, articles, books, and other research narratives that previously seemed daunting. I find the satisfaction of seeing what I accomplished over time to be rewarding, and over the long term, this helps me sustain a positive time attitude. In addition, I periodically reward myself for an extended period of productivity (or for a particularly productive week). It is easy to keep track of progress, and without exception, week by week, my writing routine leads to higher productivity than I would have accomplished without monitoring.

It is easy to establish a similar reading routine by tallying when a source has been read and annotated day by day. If one is working on an annotated bibliography, a useful monitoring approach would be to tally the numbers of sources read and word count of annotations. Keeping track of reading and writing progress like this can then be considered when assessing short-term goals from previous weeks and planning goals for the next week. Monitoring is a great way to support a weekly reading and writing routine.

Prioritization and Monitoring Exercise

Try this exercise for a two-week period, and if it works, continue it after that period. At the beginning of each week, delineate three to four tasks that you would like to accomplish by the end of the week. Be realistic: do you wish to read an article or chapter per workday and to annotate each one (this would be five sources read and annotated)? Does your schedule allow for that commitment, or should it be four or perhaps three sources? Reading sources counts as one task, and annotating them counts as another. What other research-related tasks need to be accomplished each week? Do you need to schedule and have a meeting with your major professor? This represents another task. Is there any departmental, college, or university degree paperwork that is due? Is there a data source that you would like to explore? Once you have delineated three to four of these tasks, you have scheduled your week. Make sure you write your tasks down either on a grease board or in a dedicated notebook.

Next, inspect your schedule and choose a limited period each day to dedicate to your reading and writing routine. Choose carefully; make sure the window of time is one in which you are likely to be rested. Also, make sure the time slots are ones in which you can find a quiet place to work undisturbed. Each day during the allotted time, work only on your reading and writing tasks. I mentor my students not to populate their reading and writing periods with routine classwork (which is, anything that does not relate to writing for their own research) or other tasks that do not relate to the unstructured learning challenge of becoming a researcher.

To monitor your progress use either a grease board, a notebook, or a spreadsheet. Each day, keep track of the word count that you wrote and how much reading you did. It may be that you were not able to read an entire article, chapter, or section of a book. However, this will depend on the type of reading you are doing. You may, for example, move through a whole source or a couple of sources if you are inspecting it to see if it is an important one for your research. In such cases, your reading might focus on the introduction, discussion, and conclusion sections of the article or chapter; you might skim the methods and results. If it turns out that the source is an important one, make sure you state as much in your annotation; you will later return for a deeper reading. As you monitor, demarcate whether or not you were giving the source a shallow or deep read; that way when you look back at your week you will be able to see that deep reading of a couple of sources represents significant progress as does shallow inspection of a higher number of sources. You might also keep track of the number of important sources that have been deeply read and fully annotated.

The point of this exercise is to have two weeks as samples of time spent on reading and writing that you can then look back upon to see a record of your productivity. At the end of each week, take a few minutes to compose a short journal entry reflecting upon whether or not your routine and your monitoring is helpful for motivation—and, of course keep track of the word count of your journal writing!

A final aspect of this exercise that is critically important is to take the time to share progress within a social network. This could be accomplished in a number of ways. Ideally, students in a cohort should share their progress in a weekly meeting. If there is not a cadre of students, perhaps find a student partner in your field or a related one. If one is not surrounded by a supportive cohort or student partner, then share progress with your major professor. If your major professor is not supportive, this may be a sign of bigger problems, such as a need to find a different mentor.

Summary

This content of this chapter hinges upon two exercises. The first exercise concerns time awareness, which provides a basis for embracing a reading and writing routine. Further, the first exercise clarifies the importance of task delineation for short term planning, which should be nested within long range planning. A problem with independent research, not to mention the uncomfortable educational challenge of becoming a graduate student in an unstructured learning environment, is that milestones of productivity are few and far between. Previously, during the undergraduate career, students could rely upon a structured experience designed into courses and semesters to stay motivated. Although the graduate student takes courses, thesis research expectations are unstructured. Thus, short term planning and establishing reading and writing routines are critical; as the student becomes an independent researcher, she or he learns to produce her or his own structured environments with intermittent goals and milestones. The annotated bibliography exercise from Chapter 1 provides the content with which to fill one's new routine; however, it will be difficult to stay motivated unless the student maintains awareness of progress. The monitoring exercise is important because it provides an important pathway for staying motivated. It is important to recognize progress on research between major milestones, such as completion of the thesis proposal or thesis. In the end, it is all about establishing and maintaining a positive time attitude.

Chapter 3: The Research Topic

The annotated bibliography represents only the beginning of the transition to becoming an independent researcher. Establishing the time management habits of a reading and writing routine helps one pave a pathway toward independent research; however, the real challenge requires building a foundation in one's field, developing an ability to identify research topics—particularly ones that relate to gaps in knowledge—and articulating the purpose of one's research through voicing questions. As the researcher heads down this path, he or she increasingly focuses on research objectives that will culminate in a thesis. Sounds straightforward, doesn't it? To the contrary, it is a difficult transition for most graduate students. I first began thinking about this transition when I taught the undergraduate capstone course in my department, which (at the time) was supposed to help the student develop and complete a research project.

Having worked with entering graduate students who clearly were not prepared to do independent research, I did not believe that senior undergraduates could do so either. When asked early in the semester, my students in that class would confidently portray themselves as entry-level researchers. I privately wondered, "Can they even tell me any anything about the field they profess to be part of?" Thus, I put the research project aside and started with an assignment that I call "The Interest Essay" in which students were to describe their field and to identify and summarize a number of topics that are typically researched within it. The work that students turned in portrayed that the problem was much worse than I previously recognized. Despite having taken courses such as Economic Geography, Geomorphology, Hydrology, Meteorology, and other upper division electives that cover the basics of different fields, most students could not identify their interests, responding with, "What do you mean?" Even, when prompted with, "What upper division electives have you taken?" most students were not able to recognize the subjects of those courses as fields of research. As a result, I retooled that first assignment to require that students simply identify several research fields within their major. Neither the students nor I found the process enjoyable, but I asked them to lean into it. We were all becoming aware of an educational myth that students and faculty are often complicit in producing, the myth that students are experts, which piggybacks on the false belief that becoming a researcher is easy.

The Myth of Expertise

There were intelligent students in that capstone course, but they clearly knew nothing about independent research. By the end of the semester, we moved together from identifying fields to writing The Interest Essay, to eventually identifying gaps in knowledge, and proposing research questions with related objectives. I vividly recall one student meeting with me after the semester stating that he "felt betrayed" at the end of his undergraduate education, saying that he had been "led to believe" that he was becoming an expert. He wondered why the subject matter of the capstone was not offered earlier in the curriculum so that students like him could develop their interests during their junior and senior years. I wondered the same thing, and our faculty created just that course for freshmen and sophomore majors the following year.

At the beginning of that semester, there had been an unspoken agreement that if students did well in their upper division courses, they were becoming experts. There were incentives to encouraging this myth; faculty members must increasingly concern themselves with maintaining enrollments in courses under the pervasive corporate university model. Similarly, it is more common that students simply want a course behind them so that they can claim credit and "get their degree" than it is for them to desire to learn about a field in an upper division elective course. There is no doubt that "getting a degree" is incentivized in contemporary American society. This is not news; indeed, it summarizes what many faculty members are frustrated with in the classroom. The point is that an undergraduate education that supports this myth is not conducive to producing independent researchers. It is no surprise that entering graduate students may not be able to write about research in their field; master's students in their first semester, when asked to identify a topic, commonly choose something very general, such as "global warming" or "social justice." Alternatively, they may describe their topic as a field, such as water resource management or political ecology. They often resist criticism that such "topics" are unfocused; indeed, that they have much to learn can come as a shock.

In the unspoken agreement of undergraduate education, courses are commonly highly structured so that students know what to expect, can meet the requirements, and can earn credit (and a grade). Personal habits that underlie a structured learning environment—for example, studying on days before tests or writing term papers at the end of the semester—are those that make the unstructured challenge of independent research very uncomfortable. Independent research, in contrast, is all about establishing good habits that are part of a consistent routine. As a result, the annotated bibliography assignment is only a start to the transition to becoming a graduate researcher. The next step in the transition is learning to write about one's field so that practitioners would recognize its accuracy. The annotated bibliography provides a critical foundation for writing the Interest Essay, which represents a first step toward zooming in toward one's thesis research.

The Interest Essay Exercise

This exercise is to write an essay that will provide your reader with an introduction to the research field of interest to you (see Appendix 1, Exercise 4). This statement should provide a definitive description of the field (e.g., Medical Geography, Ethnoecology, Community Ecology, Industrial Ecology, Geoarchaeology, Historical Ecology, or other areas). Disciplines such as Political Science, Human Geography, and Archaeology are too broad to count as fields in this essay because they comprise a number of focused research areas; so, choose a field (or subfield) within these broader disciplines. The statement should answer: why are you interested in a particular field? Further, it should introduce several research topics within the field that are of current intrigue and state why they are important. At the end of the essay, choose a topic that you will develop further and state why it is of particular interest to justify your choice.

To introduce the field, first try to offer a definition of it, such as "archaeobotany is the study of plant remains from archaeological sites to answer questions about the conditions of past human diet and past landscapes." Next, describe the field in more detail. To do so, you must be able to answer a number of questions: 1) What is the subject matter that researchers focus on? 2) What types of data do researchers work with? 3) What types of methodological approaches are commonly employed in the field? 4) What are some important implications of the research? To answer the last question, one must be able to envision what understanding and intellectual products would be missing if research in the field was not being done.

Notice that after introducing the field, the exercise turns toward describing particular research topics within the field. *The Craft of Research* by Booth et al. (2008:36) is an excellent resource for learning about topics; the authors state that "a research topic is an interest stated specifically enough for you to imagine becoming a local expert on it." Thus, within a field, researchers focus their efforts upon areas in which they become experts; they ask questions and assemble data as evidence to support answers to those questions when they write about their research. Researchers publish articles, book chapters, and books about research questions on the topics they study. In the Interest Essay, describe multiple topics that researchers focus upon in the field. For example, within the field of community ecology, researchers might become experts on species composition of food webs within fish communities; alternatively, another community ecologist might focus upon the influence of abiotic nutrients on energy dynamics within a food web. Still other researchers might study how communities interact across the landscape as metacommunities and determine what processes influence connectivity and fragmentation over time. The Interest Essay, then, should enable the student researcher to reflect upon the topics that were of interest to authors of sources that he or she summarized in the Annotated Bibliography Exercise. The student should be asking, how can I describe the topic of the research presented by the author?

The Interest Essay represents a step beyond the annotated bibliography toward focusing one's research interests. This is a critical element of the exercise; if the student treats the Interest Essay as an assignment like an average undergraduate student would, he or she might not pay careful attention to which three or four topics are described. But the point of the essay is not simply to finish it; rather, the goal of the exercise is to grow more confident about narrowing down one's interests. Thus, choose examples of research within the field that are particularly interesting and describe what topics they cover. If there are multiple sources that can be described very similarly, then they likely represent research on the same topic. Indeed, an excellent way to describe a research topic is to articulate what it is generally and then provide a real-world example from the sources one has read. For example:

> Metacommunity analysis is an important topic within community ecology because it focuses upon factors that connect or fragment animal communities across the landscape. For example, Driver (2015; Driver and Hoeinghaus 2016) in his doctoral dissertation studied the influence of stream flow in intermittent streams in north Texas on connectivity of upstream and downstream fish communities. He found that long periods of drought led to decreases in fish community diversity, which could only be re-established once upstream and downstream communities were reconnected as a metacommunity by rain events that reconnected stream flow.

The uninitiated reader might have questions about this research, such as, "What is meant by fish community diversity?" Or, "Why did diversity decline with drought?" Diversity in this context means the number of fish species interacting in a portion (length) of the stream. When portions of the stream were disconnected due to limited flow and drought, stream habitat gradually declined and increasingly sensitive species died off in those reaches of the stream. Downstream, there are larger bodies of water where many species survive; once a drought ends, the stream can become reconnected and those species have the opportunity to migrate upstream, increasing upstream community diversity. Driver's research concerned whether or not (and if so, to what degree) such community diversity changes occur.

Note, that I do not do fish community ecology research; yet, I took the time to describe metacommunity ecology as a topic. To do so, I had to empathize with the audience I was trying to reach by answering questions I suspected they would have. When doing the exercise, describe each topic in similar fashion. Near the end of the essay, select the topic that is of greatest interest; this will become the subject of the next exercise, The Topic Essay, which is to write about several research questions that can be asked within a topic. Prior to focusing on research questions, and before diving deeper into becoming a researcher, it is important to discuss a critically important aspect of research in more detail. What is the audience for whom one writes?

The Audience

Entry-level researchers typically only have undergraduate writing experience, which rarely concerns summarizing independent research. Writing during college takes place in composition classes, essay exams, and term papers. Each of these draws upon summarizing previous work; alternatively, composition may require the student to reflect upon their own interests.

The results tend to be third-person summaries, reports that answer structured questions, or first person narratives that are reflective. The audience of such writing can be the professor, the writer, and at times other students; however, the audience is not practitioners of a field of research. In my experience, it is difficult for new graduate students to write about why a topic or field is interesting because as a non-expert they feel they have no authority to do so.

Unsurprisingly, students often default to the writing style they learned during their undergraduate education. For example, some students write the Interest Essay as a first person reflective piece, stating at length why they find a field or topic of great personal interest. Following this flawed strategy, students go to great lengths to describe their motivations, which might include personal histories, anecdotal stories, descriptions of undergraduate classes, and other personal experiences. The literature and theoretical review of interest is not about the student, but about the topic of the student's interest—the existing theory, research and what needs to be done. Only briefly should one write in the first person that he or she intends to pursue research to fill a gap in knowledge. As an exception, most social scientists collect data by interacting with research subjects. Here, writing in the first person about one's own findings is appropriate and expected. Social scientists still ensure that their *research subjects*, not themselves, literally remain the subject of the research. And discussion of the existing research in the field is in the third person. The critical point here is that students cannot assume that members of their audience will be interested in their motivation; such is generally not the case. Members of a research community want to know why *they* should be interested in a topic, not why the student is. Thus, the first person narrative strategy, if relied on too heavily, can represent an early pitfall in research writing for graduate students.

The second undergraduate writing approach is to provide a dry, pedestrian summary of previous works on topics in the field. By adopting this strategy, the student writer has erred in the opposite direction from the first-person approach. By not voicing why other researchers should take an interest, the writer is essentially stating, "You're an expert; figure it out yourself." Alternatively, because the student is not yet an authority, they may be thinking that it is not their place to state why a field or topic is of interest to researchers. However, the audience is the community of researchers, and the job of research writing is to make reading worth the time and effort of members of that community. To trust that something is valuable, researchers in the field need to develop trust in the writer's assessment of a field or topic. It is important to note that beginning with the annotated bibliography is a critically important stage before writing about one's interests. It is through extensive reading of sources that students become competent and confident enough to portray why a topic is of interest to members of the research community.

A simple, direct approach is necessary for engaging writing in the third person. First, write the dry summary of an article for the annotated bibliography. Second, however, ask and answer, "What is interesting about an article, a topic, or a field in general?" Third, practice voicing thoughts in the third person like an expert would: "surface water hydrology in alpine contexts is important because of… and is of interest because…" As your research writing skills develop, you will find that dry summaries become progressively easier to fix. A particular fact or idea may require a pedestrian description, but the paragraph can be improved by book-ending that description with a strong introductory sentence that states the importance of the research and an engaging concluding sentence that highlights why the research is interesting. These subtle shifts in writing style keep the reader engaged because you will have gained their trust. Learning to write for the appropriate audience—the research community in one's field—is a critical step toward becoming an independent researcher. Gaining that skill makes it easier to stay motivated and helps students take the next couple of steps, which are to write about topics and then about research questions tied to research plans and objectives.

The Topic Essay Exercise

A working definition of "topic" from Booth et al. (2008:36) is a specifically stated interest on which one is capable of becoming an expert. This is narrower than a field of interest. One might, for example, claim to be a cultural geographer, which is a field of research with shared interests, concepts, methods, and principles, that focuses on human interactions with landscapes. There are, however, a large variety of topics that one can study within cultural geography, such as the historical and contemporary influences of land modification (e.g., terracing, reservoir building, land clearing, etc.) in, say, central Mexico. Inexperienced researchers who have not committed to diving into a topic often claim expertise in a research field. This represents nothing more than the myth of expertise discussed earlier and instead merely reflects an initial curiosity about subject matter but not the acquisition of skills, knowledge, and experience to do research within that field. To become an expert, one must settle on a narrower topic to situate concepts and apply methods of the field in a particular context. Indeed, it is the combination of previous knowledge of the field and an unstudied context (or case or example) that represents a gap in knowledge.

Taking Booth et al.'s definition a step farther, we can employ their advice on how to narrow one's curiosity toward a meaningful research topic. First, one must know the research field well enough to recognize where gaps in knowledge lie. Second, a student must master a working knowledge of the field to understand what it means to become a research practitioner—that is, she or he must identify the field's common methods, theories, and approaches. The Annotated Bibliography and Interest Essay exercises will help the beginning researcher establish this foundation. Third, a researcher can then ask a series of direct questions about her or his interests that help identify and narrow down a topic. What will be my subject of study? Is it people, ecological communities, artifacts from ancient sites, historic documents, biological tissues, or another subject? Second, where will be my context for the research? Will my research focus on a subject within a particular location, across a region, between one or more communities? Alternatively, will the research be non-spatial but experimental? Finally, how will I do the research? What methods will I employ and why will I choose them? I have voiced each of these questions in the first person, which is appropriate when considering one's own research. The step of zooming in upon a topic, however, is quite difficult for beginning researchers, and an important means to gain experience is to write an Interest Essay (summarized earlier in this chapter) about previous research by experts in the field. When studying others' work, ask the same questions about their research. A major milestone is to commit to one's own research topic and to answer those questions concerning one's own interests, which is the purpose of the Topic Essay (see Appendix 1, Exercise 5).

The Topic Essay[1] will provide your reader with an introduction to the topic you are choosing to develop. This statement should provide a definitive description of the topic within the research area. The statement should answer: why is the topic important in the research area (or why does it matter, or to follow Booth et al. [2008:47, 56] "so what?!")? It should briefly contextualize the topic into related, previous research by answering questions such as: Has there been research on the same topic in other places? Has previous research been done in the context of interest (e.g., within the same population, community, species, region, or other type of context) using a different methodology or theoretical framework? Further, the Topic Essay should introduce multiple research questions that relate to the topic either to solve practical problems or to increase understanding related to conceptual problems. At the end of the essay, choose a research question (or multiple questions) that you will develop further and state why it is of particular interest to justify your choice.

Briefly, a conceptual problem is addressed when a pure research question is answered (Booth et al. 2008:59). Pure research relates directly to gaining of understanding within the field of interest. The cost[2] of not doing such research is that there will be no increase in understanding. In contrast, applied research has practical implications; the cost of not doing applied research is no progress toward solutions to real world problems. Research can have pure and applied implications if increased scholarly understanding also provides solutions to practical problems. A geneticist may be interested in the implications of genetic modification upon the evolution of a species, but the modification could also have practical implications for pest management, medicine, or other areas of applied research. The purpose of the Topic Essay is to provide a summary of the importance of the topic; conveying important research questions and stating the consequences of leaving them unanswered is a straightforward way to convince the reader of the importance of a research topic. Clearly, a consequence relates directly to the gap in knowledge that meaningful research on a topic will fill. Moreover, articulating the consequences highlights the reason it is important to fill that gap.

---

1 Many of the concepts used to frame topics in this essay exercise are directly from Booth et al. (2008, Chapter 3).

2 Costs and consequences related to applied and pure research problems are covered in detail in Booth et al. (2008, Chapter 4, pages 53–61).

Topic Ideas from the Major Professor

When a student begins her/his graduate career, she/he does not have extensive experience evaluating a field or topics within it. A critical component of the master's degree is learning the skills of research, and though it is important to choose a topic that is of interest, *the actual commitment is more important than the topic itself.* With the obvious caveat that students should work within their field of interest, I routinely tell my graduate students that any topic is better than no topic. It is more important to settle on a topic that fills a gap in knowledge and that lends itself to a feasible research plan during the two-year window of a master's program than it is to make certain that an ideal or perfect topic is selected. If a student is serious about becoming a researcher, she or he will confront numerous research projects within the field during her or his career; that is, there will be plenty of opportunities to focus on a diversity of topics in the future. The student, thus, may be caught in a bind; they are establishing the skillsets necessary for identifying a suitable topic, but they also need to make progress on developing research skills within the field during the short period of a master's program.

The major professor (a.k.a. graduate advisor, thesis chair), in contrast, has considerably greater experience in the field and will have selected, developed, and completed a number of research projects. Further, they went through the same process that the student is beginning. As a result, the major professor is an excellent person to turn to for ideas concerning topics. Students sometimes balk at this idea, conveying, "Then I wouldn't actually be doing the work, and it would not be *my* research." Students may cling to this perspective stubbornly, as if a purest approach to graduate school is wise. It may not be wise for several reasons. First, whether or not the student selects the topic, the major professor is likely to be deeply involved in the research and will probably be a co-author on any scholarly presentations and publications out of the work. Graduate students (particularly master's students) often become a member of a collaborative team. The reason for this is simple, at the beginning of their research program, the student does not yet have the skillsets to pursue the research, and he/she requires the experience and support of others. Sometimes students dismiss this reality, which comes off as nothing other than hubris to their major professors and committee members. My reaction to such dismissal is that it represents the myth of expertise discussed above, and I am not willing to pretend that my students are experts if they are unable to select a topic that leads to a feasible research plan. If they are unable to do so, either they must be open to receiving my guidance, or I will not continue to work with them. I usually assess whether or not the process will work by the end of the first semester of the program, and I interview each applicant who wants to work with me to gauge whether or not they seem open to a collaborative working relationship. In short, I have little patience for the myth of expertise because mentoring graduate students is a serious commitment, and we all have new things to learn. I am not unusual: You should expect a similar disposition from your own academic advisors.

Second, the major professor may have a number of topics in mind and have envisioned road maps for how to pursue the research. If willing, the student can benefit from that experience by shortening the time and effort for selecting their topic. Together, the student and major professor can discuss components of the research in a stepwise manner. The major professor will likely have a number of sources in mind that the student should begin reading, and the student can gain ownership by starting an annotated bibliography with those sources, by establishing a bibliographic trail of her/his own, and by bringing new sources to the major professor's attention. During my graduate education, this became somewhat of a contest. My major professor had exceptional command of the literature in my field, and I would try to find new sources that he had not encountered. Not only was my major professor's knowledge extensive, he had a great memory for locating hard copies of sources in his personal library. Many was the time that I would arrive with a fresh article, thinking I had finally bested him only to have the response be "third file cabinet, top drawer, three-quarters of the way back," where I would find his copy of the same article.

There are other advantages to working closely with the major professor during topic selection. The student can more easily begin to write about their field using the Interest Essay because they have a topic in mind. They can look for related topics, and will be more likely to recognize the kinds of research questions addressed in the field. Further, narrowing the focus to a topic not only takes advantage of the mentor's experience, it exposes the student to the methodological approaches and theoretical foundations they will need to learn along the way. The major professor wants to see her or his students grow as researchers. Mentors become excited when students bring something new to the table. If a student is willing to take the major professor's advice early in the research process and to accept help with topic selection and a research plan, a dialogue will develop. This mentoring relationship then leads to sharing of research experiences, something that the student must develop to succeed.

Summary

It is exciting for students to enter a graduate program. Acceptance to a good program means that faculty members have reviewed the student's portfolio and decided that she/he has potential to become a researcher. This represents an investment by the faculty in the student. As described in this chapter and in previous ones, the undergraduate education system tends not to prepare students to be researchers. Rather, they will have developed fundamental learning skills. Since good graduate programs tend to select the best undergraduates for admission, it is easy to support the myth that entering graduate students are already experts in the field. They are not; in fact, that is what graduate school is for—to become an expert through learning to be an independent researcher. As a student, you will first dive into a large literature, perhaps aimlessly. It is natural to be frustrated early; you will quickly learn that not having a research topic with a clear research plan is a disadvantage. A critical step is to fashion a working relationship with a mentor. A great way to start this relationship is to ask for advice on what to begin reading while beginning the annotated bibliography exercise from the previous chapter. If you follow your mentor's advice closely and ask for more, eventually you will move into thinking about your field more holistically; the Interest Essay challenges you to write about the field, but also to identify, summarize, and state the importance of multiple topics. Engage your major professor in conversations about these topics. Tell them why you think the sources you are reading are important. In the process, try to identify gaps in knowledge in the field. You are not yet an expert, and it is likely that you will feel that you do not have handle on the literature. You may ask, "But I am just starting, I don't know anything... how can I figure out what topic to select?!" This is likely to be stressful because you will be taking courses, perhaps assisting professors with teaching, and you will be hearing from faculty that you should be "making progress on your research."

Do not despair. If you have been working on your annotated bibliography, writing about your field, and regularly talking with your major professor, you will have gained their respect. They may have a few topics in mind, and you should seek their advice. The best way to settle on a topic is through mentoring. Once you have a topic, learn more about it. Write about it using the framework of the Topic Essay, and seek feedback from your mentor on what you have written. Once you have gotten to this point, you will no longer require a structured learning environment to succeed. Congratulations, you are now becoming a researcher!

Chapter 4: The Support Network

If you decide to follow the advice presented in the preceding chapters, it is likely that the process of a becoming a researcher will be frustrating (at least part of the time, if not more often). Several elements of learning independent research are challenging:

1. the transition to an unstructured learning process may be confusing and disorienting;
2. one begins the master's research process before knowing their topic, making it difficult to choose what to read, what to write about, and what to focus on;
3. the process of narrowing down to a topic requires a commitment to something that is relatively unknown in an environment that demands high performance with knowledge work;
4. gaining focus and confidence seems agonizingly slow; and
5. your work is being reviewed and criticized while you are trying establish your footing in your field.

There are additional parts of the transition that are frustrating, but these are some important ones. If you are in a program that challenges you to become an independent researcher, your experience will be much different than it was during undergraduate education. Not even your coursework will feel the same because it is likely that you will take classes oriented toward preparing for thesis research.

Assignments in Chapters 1 through 3 are structured to begin working toward becoming an independent researcher despite (or to overcome) the likely considerable discomfort students face during their graduate school transition. The process starts wide by reading and writing about sources just to get a handle on the field—to simply begin—and narrows to writing about topics. As one moves from annotating sources to writing about their field to selecting and writing about topics that interest them, she or he begins to triangulate toward selection of her or his own research topic. It is tempting to try to tackle this transition alone; after all, the purpose is to learn to do independent research. Even if a student establishes an effective reading and writing routine, however, she or he will not succeed in isolation. The main reason is that at some point, research will be shared with an audience. Three support networks prepare the student for becoming an independent researcher who can confidently and effectively share the products of their research: the mentoring relationship with the major professor; supplementary mentoring from the thesis committee; and one's student peer cohort.

Relating to the Major Professor

Graduate students should have a close mentoring relationship with their major professor. Establishing that relationship varies from easy to difficult because personalities are diverse. What are some expectations of that relationship? What are ways to maintain a healthy relationship with the major professor?

Students can expect that major professors will admire them when they take advice. Take the annotated bibliography, for example; if a student requests five to ten sources that are important, the major professor will simply expect the student to read them carefully within a few weeks of that conversation. In addition, mentors expect honest self-appraisal by graduate students. If a student reads a source and is confused, then they should ask questions. Often students are worried about appearing confused or incapable, but major professors can only provide guidance if there is an open dialogue.

If a student does what is advised and needs help, she or he should expect to receive help. The major professor-student relationship, however, is not one that comes with the structured norms of undergraduate, classroom education. It should be treated as a professional relationship. To establish a reliable and healthy dialogue, value the major professor's time. As Davis and Parker (1997:9) note, assume that "faculty are a scarce resource." This presents a tricky challenge, one that if addressed poorly by the student can cause many problems. If a student seeks help each time there is a question, for example, it will be difficult to respect the major professor's time. In addition, the faculty member is likely to conclude that the student does not have the ability to work independently. Here are some tips for respecting faculty time:

1. Jot down questions about sources in your annotated bibliography in a notebook or document. Bank up several questions to bring to the major professor. This has the added advantage of counting as research writing and is particularly valuable early in graduate school when one is setting his or her routines. With continued reading, you may also answer many of your questions.
2. Seek a regular, reliable, agreed-upon individual meeting time with your major professor. You may win the lottery and begin your program with a major professor who approaches you about setting these meetings and invites them. Professors are human, however, and may be introverted, disorganized, unreliable, and unwelcoming. You need their expertise, and making your time with them reliable will work with most major professors. If not, that person might not be a good mentor.
3. Have an agenda for each of those meetings, one that includes time for your questions. Make sure you have no more than five agenda items.
4. Prioritize the agenda items in case you cannot get to everything. That way the most important stuff gets covered first.

5. Actively read so that you accumulate a bank of questions for conversations with your major professor.
6. Stay dedicated to the meeting routine. Even if there is little to cover, make sure the meeting time stays available and that you touch base to keep the dialogue open. The major professor wants you to be independent, so if you are never around, they will conclude that is what you are doing—being independent.

As mentioned in the previous chapter, the major professor becomes particularly important during topic selection. The major professor is likely to see the bigger implications of any particular topic for the field. As a result, they understand the context of the research while the student is still learning the skills of research. The major professor can keep the student grounded in that context even when it is difficult to see, which is likely to happen as students learn methodology, become steeped in their field's literature, and begin writing components of their thesis proposal. In this sense, the major professor provides an important anchor for the student's interests.

An important question is, "What should I do if I do not get along with my major professor?" Before answering this question, it is important to ask additional ones. First, what kind of problem are you having? If the issues are severe, such as harassment or abuse, which (unfortunately) occur in higher education, then reach out to student services at your university right away. Many problems, however, are not so severe. In those cases, ask yourself, "Have I worked to establish a relationship with my major professor (as described above)?" Also, "Is this a problem that relates to poor communication?" If the answer to either of those question is "yes," then establishing a healthier meeting routine may solve the problems. If you find that your research interests are moving away from those of the major professor, this too requires communication. A good mentor will seek to help you find the right fit for your interests. He or she may still be able to guide your research if your interests change, but to do so will require clear communication.

Students can expect their relationship with their major professor to be challenging at particular points in the process. Your papers, proposals, conference abstracts, chapters, and other documents will be criticized, which is potentially uncomfortable. If you have a reliable, professional relationship with your major professor, you can talk about their comments; often, clarification and additional communication depersonalizes criticism, which makes the writing process more effective (and more enjoyable).

Creating an effective relationship with your major professor will also be helpful when other problems arise. You may not even like your major professor, but it is the professional relationship that matters most. Additionally, your major professor will help you form your thesis committee and will be able to help you address their criticisms when they read your work.

The Thesis Committee

Students will form similar mentoring relationships with members of their thesis committee. Master's thesis committees typically comprise two faculty members in addition to the major professor. Doctoral dissertation committees typically add one or two additional members. The most effective way to form the committee is to seek guidance from the major professor during the process of refining one's interests and selecting a topic. Members of the committee might vary in their roles. Some thesis committee members will provide additional close mentoring if their expertise is required for pursuing the research. Such members may hold skills that are different from those of the major professor. In addition, committee members play an important role in assessing the impacts and merit of the thesis research. In contrast to the major professor, however, students usually meet less frequently with the other committee members.

Three important milestones involve the thesis committee (Appendix 2, Example of Student Handbook). The first is planning coursework (see Appendix 1, Exercise 6). Classes taken during the first semester might be foundation courses required for the degree; electives should be chosen carefully with guidance from the major professor. Once a student has formed the committee, its members may have advice on courses to take. Alternatively, the student may want to seek out advise from those faculty members to make sure their classes support conceptual and skillset development that relate to the research. The major professor and other committee members represent an excellent resource for designing one's degree course plan.

Second, the first year of graduate school (particularly for the master's degree) is oriented toward the process of topic selection and framing research objectives. The exercises presented in this book guide the student through that process. By the end of the second semester, the student should be well on his or her way toward completion of a research proposal. The details of what that proposal will look like are covered later in this book, but the thesis committee will review that proposal to determine if the objectives are meaningful and if the research plan is feasible. Many programs require a formal proposal defense, which involves providing a public lecture on the research followed by a closed-door thesis committee examination. In the context of a thesis proposal, "public" means open to students and faculty members in one's program and related programs.

Proposal review is a critical step in the process of becoming a researcher. Critical review of independent research is essential; however, it is also uncomfortable for students who are not used to the unstructured learning environment of research. In classes, students get grades and perhaps a few comments on their work, a process that ends once the assignment is graded. The thesis, however, is a living document that develops out of the proposal; thus, criticism from the committee members promotes improvement of the research through revision. The committee will require revisions of the proposal until they are convinced that the student is ready to move forward with the research. This an important reason why construction of the thesis committee should be done with guidance of the major professor. A potential committee member, for example, may have the intellectual skillset that is needed, but may be difficult to work with. The major professor can help the student anticipate any problems and navigate the process of review.

The third milestone that involves the thesis committee is the thesis defense. Writing of the master's thesis takes place during the second year (the process occurs later for the doctoral dissertation); the major professor reviews and edits the document extensively. Once the major professor is satisfied with the thesis, a defense draft will be circulated to the committee members for review. The thesis defense consists of a public presentation detailing the topic and research objectives, communicating the methods and analyses of the project, and summarizing the outcomes of the thesis. This is usually followed by a question-answer period for the public members of the audience (usually other students and faculty members from the student's program), which is followed by a closed-door examination with the committee members that focuses on questions about the research. If needed, the committee will require the student to revise before accepting the thesis.

How students interact with thesis committee members is important. Committee members are likely to have students of their own whose research is their primary focus. Thus, the feedback of committee members is a scarce resource, even more so than that of the major professor. It is important to be cognizant of their role in the proposal and thesis process, and to be thoughtful in making plans for proposal and thesis review. That said, thesis committee members are obligated to review proposals and theses in a timely manner, commonly within two to three weeks of receiving the document (expectations may vary by program, so find out). When you send them a document for review, politely request a target date for when you will receive their feedback. A good strategy is for the student to politely remind the committee member of those targets the week prior to the agreed time. If a thesis committee member is uncommunicative or exceptionally late, ask for help from the major professor in reaching out to that committee member. With planning and thoughtful communication, however, the proposal and thesis review processes usually go well. Faculty members want students to succeed. A common rough spot concerns student handling of deadlines.

Destigmatizing Deadlines

In the realm of time management, deadlines can be your best friend or worst enemy, particularly in interactions with your major professor and committee members. Left unattended (by me as their mentor), my graduate students commonly treat deadlines in one of two unhealthy manners. The first response is to set a deadline (which I typically refer to as targets or target dates) and to stick to it at all costs to the point of extreme stress and anxiety. A target date is only useful if it provides the structure to finish work, such as writing a section of a proposal or chapter, in a reasonable amount of time. In the context of this book, "reasonable" can be described as "achievable within the constraints of the writing or reading routine over a defined period." An important contingency of research is that one often learns during the process of writing, and thus the goals of writing might change. It may be that a target date needs to be amended in order to finish work related one or another goal. In addition, life happens! There are emergencies, illnesses, and tragedies in life that disrupt plans. It may seem obvious that target dates should change under such circumstances, but many of my students increase the pressure to meet a deadline when tragedy strikes. Such habits completely undermine an ability to set a routine that supports a reasonable, effective, and achievable research career. The target date is set to provide structure; its purpose is to have something towards which to work. *It is rarely a date after which failure happens.*

The other common response in my students is either to avoid setting target dates for tasks or to ignore them once they are set. This tends to relate closely to a general problem with committing to a routine. It is not actually the goal that such students tend to avoid but rather the routine. Setting a target date represents an impending reminder that a routine has not been set. To solve this problem, the student must realign her or his time management goals in the context of reflecting upon their aspirations to become a researcher. If establishing a routine is an insurmountable challenge, then work is likely to be delayed or to remain unfinished. The major professor may eventually believe that the student is not truly interested in becoming a researcher. Setting and keeping a reading and writing routine is the best way to overcome fear of deadlines because even when an extension is needed, the student knows she or he has the routine to rely upon. Even if work is going slower than expected, new (reasonable) expectations for completion of tasks can be set *because one knows that work is being done.*

It is important to realize when others are relying on completion of work on a deadline. Collaboration with others is one reason students (and other researchers) become overly anxious about meeting targets. Regardless if others are involved in the work, contingencies on time can still become problematic. There are ways to offset additional stress caused by missing a deadline. Send periodic progress updates to collaborators. This enables partners in research to envision the pace toward completion of tasks. In progress updates, state how far along you are toward completion and review what needs to be finished. Conclude the update with an appraisal of whether you think you will meet the target date. As you near the target date, it should be progressively easier to envision if the deadline will be met. That way, if an extension is necessary, your request for one will not catch collaborators by surprise. Indeed, given that you have been a good communicator, collaborators are likely to be supportive of the extension.

One additional note regarding deadlines concerns the importance of cohort and major professor meetings. In my research group, we set tasks each week publicly on a grease board (see Figure 1). Each meeting, when we review work from the previous week, we talk about what went right and what went wrong in terms of task completion. In the process, success and failure at meeting deadlines become normalized; that is, progress becomes routine conversation. Success in research is all about the routine!

The Cohort

From the day that a student enters a program to the day of graduation, he or she is surrounded by other students who entered the program when they did. Students who begin graduate studies together are part of a cohort, a group that matriculates together. That group can be either a cause of dissension or a source of support. Since personalities vary within a cohort, one can expect that some students will need more support than others and some students will be more supportive of their colleagues than others will. How individuals treat each other within the cohort really matters, because a group with a supportive culture can dramatically improve not just the comfort level of students but also the quality of education for individuals and for the program. Members of a cohort represent the brand of a program; working together they may become collaborators. A cooperative cohort will have wiser graduates than one that focuses on competition at the expense of cooperation. There are ways to build a cohort, and there are behaviors that can ruin a cohort, which should be avoided.

Behaviors to Avoid

The competitive admissions process for good graduate programs may lead you to think that program environments should also be highly competitive. There are at least two kinds of competitiveness, however—winner take all versus cooperative competition. The first position will not serve you well if you hope to develop professional habits that help you become part of a research community. Overly competitive graduate students are divisive; they exhibit behaviors such as argumentativeness, arrogance, dismissiveness, and hypercriticism of others. As a result, such students stand out like a sore thumb. Indicative behaviors include routine attempts to dominate seminar discussions to try to impress professors, excessive commentary on the work of student colleagues in classrooms and other public settings, and hypersensitivity to criticism from others (including professors). A good look in the mirror can help you determine whether you have any of these fatal flaws, and it is fair to say that most people—when at their worst—exhibit some of these behaviors. That said, if your look in the mirror leads you to conclude that perhaps you engage in these behaviors often, here are a couple of pieces of advice. First, it is likely that you see success in others as a threat to your own progress and failure in others as a boon to your own advancement. Simply try to turn this on its head and attempt to see something (anything) good in your colleague's work. Second, exercise restraint. Truth be told, your criticisms might be valid, but if there is a fatal flaw in someone else's work, it will eventually be corrected. And, if not, perhaps it is not as fatal as you think. You lose nothing by holding your tongue and seeing what develops; you lose a lot if you put hypercriticism on display. I know because I was one of these people early in my graduate career.

I vividly recall criticizing the ideas of one of my colleagues during my master's program. To be honest, arguing is sometimes fun, and in the heat of the moment I pushed a colleague too far. She was offended. Two things happened. First, the interaction was just prior to class, and the professor entered mid-argument. He cut me off, looked me right in the eye, and said, "You don't get to be critical of others until you put your own ideas out there for others to criticize." That stopped me in my tracks, but to be honest, at that point in time, I was most worried about what the professor thought of me. Second, I learned a more valuable lesson years afterwards, in hindsight. I regretted that I had offended someone who remains a professional colleague. When in conflict, if neither you nor your fellow graduate students change direction, you are colleagues for life: you will be peer-reviewers for each other's work, interact at professional meetings, and teach or become co-workers with each other's students. My advice to you is, learn this sooner than later so that if you make a mistake, you can redirect. You might need to apologize or you might be able to show restraint and avoid a similar mistake to the one I made. It is critical to remember that the success of student colleagues improves your own chances of success not just because of a more supportive environment but because successful graduates improve the brand of the program.

Behaviors to Embrace

One of the most important behaviors to encourage in yourself and others is to establish the reading and writing routine discussed in previous chapters. Interactions with your colleagues will either support or detract from focusing on your own learning. It is relatively easy for students to band together to establish social outings, such as a weekly happy hour. Such gatherings are important, but they can also be counterproductive if the cohort is unhealthy. There are a number of initiatives for building a healthy cohort culture. One problem that I commonly face as a mentor of graduate researchers is that I feel that they are waiting for others to provide the structure and organization of their cohort. As faculty members, we consistently think, "if only students would create certain opportunities for themselves, their experience would improve." Here are some ideas.

Establish a reading group. Find a few students with similar interests, and choose an article to review and discuss each week. Better yet, choose a book that is central to your common interests, and review it chapter by chapter from week to week. Do this without faculty input; use it as a supplementary exercise to classwork. Make sure that you annotate whatever you read, and count it toward your writing for the week. When you meet with the reading group, concentrate on a few themes from the article or chapter. Try to bring discussion points to the table that you think are important, and ask questions. Most importantly, actively listen to the perspectives of your colleagues.

Establish a peer-review group. Engage in the process of reading and commenting on each other's written work. If you adopt the structure of this book, start by reviewing a few annotations from each group member. Then move to the interest essay and the topic essay. One way to ensure the health of group culture is to change the dynamic of how reviews are requested *by asking if you can review someone else's writing*. This changes the dynamic of peer review because it allows the group members to focus upon learning how to be good reviewers, editors, and proofreaders rather than making the group culture be mainly about seeking feedback on your own work. If the group focuses on improving the review process, there is no doubt that you will receive feedback on your own work. If you end up writing narrative reviews as commentaries on each other's work, make sure you count that as part of your writing routine. In addition, pay attention to any new sources that could help you in your own work as these can then become part of your annotated bibliography. Pay attention to tone; seek to voice criticisms as questions. This is part of learning to be a constructive reviewer; all else being equal, if an author receives the same criticism from a reviewer with a constructive tone versus one with a damning tone, they are more likely to listen to the former than the latter. To learn the peer review process, consider the following exercise.

The Peer Review Exercise

This exercise involves seeking to review the work of a student in your cohort (see Appendix 1, Exercises 4, 5, and 7). It is preferable that you approach a colleague who shares research interests with you, but even if you cannot find someone with overlapping interests, this exercise can benefit you. To set up the exercise there is one practice to avoid and one premise to adopt. First, explicitly avoid thinking of yourself as an expert. You may indeed know a lot about a research area, but for practicing peer review, it is best not to engage in this exercise as an expert providing criticism. Instead, adopt the premise that you are reading your colleague's work as a member of the research audience. Your purpose will not be to gauge whether the research is correct, warranted, and/or novel. Your feedback will concern whether the piece communicates the warrant and novelty of the research effectively. Scholars practice peer review in a number of ways; I divide the process into two steps.

First, give the essay a summary read. Take notes on what the subject matter is, but avoid forming concrete impressions about the value of the essay or research. If you notice typographical errors and sentences that are unclear, this first read is a good time to record those issues. Once you have finished, compose a brief summary of your impression of the essay. There are a number of general questions you can ask, such as, "Are key concepts defined?" "Does the author cite sources accurately?" "Are sentences well structured; are there unclear sentences?" "Does the essay flow?" "Or, does the narrative within or between paragraphs jump around in ways that are disorienting for the reader?" "Are portions too wordy?" "Does the author use multiple words when fewer words would suffice?"

Second, after a break, come back to the essay and give it another careful read. During this read, think carefully about the goal of the essay. Take, for example, the Topic Essay. Your review should respond to several questions. "Do you understand what the topic is, based on the author's description?" "Is the topic situated within a clearly described research area?" "Does the topic seem too broad or too narrow?" "Has the author stated how the topic addresses a gap in knowledge within the research area?" "Has the author stated the importance of filling that knowledge gap in a way that is convincing?" "Does the author describe examples of research questions that need to be answered to fill that gap in knowledge?" "Does the essay conclude with a paragraph that summarizes the major points that were made?"

There are subtle components to this exercise that are important to recognize. Again, when you ask to review another's work, the exercise becomes about practicing the peer review process, not about getting feedback for your own work. Peer review is a form of leadership, and it is the lifeblood of scholarly research. To be certain, it can be imperfect. A good journal editor, however, will seek feedback from multiple reviewers who span expertise and perspectives about a topic in a research area. Thus, it is essential for new researchers to learn how to become effective reviewers. Second, since the author did not come to you for your expertise but you went to for practice, this removes the need to seem like an expert. As an editor (or as an author), reviews from damning authority figures are typically not helpful. Removing the burden of expertise will help you focus on the tone of your review. A highly effective strategy is to voice criticisms from the point of view of being a reader with statements such as, "As a reader, I was not convinced of your claim in paragraph X on page Y. The statement may need to be made clearer or removed." Such a statement represents constructive criticism; the more constructive your reviews, the more you help the field develop. Importantly, for each review that you give to members of your group, request feedback. Was the review helpful? Were comments constructive and clear? *Recall that the goal is for you to learn and practice peer review.*

If you decide to practice the peer review exercise in a group (see preceding section), start by reviewing each other's Interest Essays. Go through a couple of rounds of review so that your colleagues can see the changes in your work and you can see the changes in theirs. Then move on to the Topic Essay; each of these exercises has a specific purpose, so review is straightforward. Either the essay exercises fulfil their purposes, or they do not. Eventually, you will move into reviewing each other's proposals, thesis chapters, and articles. If you engage the process as a cohort, you will be versed in peer review by the time you achieve your master's degree.

Summary

In this chapter, I have challenged you to develop support networks during your transition into graduate research. By now, you are aware that the unstructured learning environment of graduate school represents a major change from undergraduate learning. It is important to meet regularly with your major professor about your progress, but it is also critical to have an agenda and to show progress at each meeting. Anticipate where your struggles lie, and seek advice on how to move forward. Early in the process, you might ask for sources to read; soon after, you might have questions about the implications of the research from those sources. As you identify gaps in knowledge, run them by your major professor who has more experience in the field. Along the way, your major professor will be implicitly gauging your ability to write, to work independently, and to take criticism. Welcome it. An important step forward is to discuss with your major professor who should be members of your thesis committee; this will become increasingly clear as you narrow down your topic. Be prepared to discuss your topic with potential committee members and to receive their feedback. For important milestones such as the thesis proposal, make sure your major professor is happy with a defense draft. Present it to your committee with target dates for the public defense; give them ample time to review it, but check in with them periodically. If you feel you might be pestering them too much or that things are languishing, check in with your major professor for advice.

Treat members of your cohort with respect. Adopt a perspective that cooperation and mutual support is more effective than isolation and competition. Find colleagues with similar interests; start a reading group. As you move forward, seek to practice peer review. Peer review is a leadership skill in that you create a supportive environment for constructive criticism. Ask your colleagues for feedback on your reviews; were your comments helpful? Engage in this process with courtesy and honesty because learning to accept and provide constructive criticism will benefit you for years to come. If you establish these practices within the first semester of graduate school, you will have a source of feedback, which will help you frame your research questions and objectives.

Chapter 5: The Research Question

One narrows general interests in a field of study to a specific topic through focused reading, annotation of important papers, and determination of gaps in knowledge. The next step for the student researcher, then, is to focus a topic toward one or a few research questions[1]. The purpose of this voicing is that questions lead to answers, which are research outcomes or products. Once students can articulate their research as a question (or as a few questions), it becomes easier to envision a pathway forward because they can frame objectives that relate to the tasks that they must accomplish to provide an answer.

What is a good research question? Quite simply, it is one that, if left unanswered, leaves a gap in knowledge. Applied research frames questions that relate to practical questions, such as what is the optimal location for a particular type of business for maximizing profitability? The cost of not answering such a question is pragmatic and direct; if unanswered, a business owner may be earning less profit than in a better location. Pure research, in contrast, relates to questions that seek to increase scholarly understanding in a field of research but that may not have practical implications. In archaeology, for example, one might wish to understand how people hunted animal populations in different environments in the past. If unanswered, the consequence is absent understanding about how people lived. A key aspect of applied and pure research questions is that something is gained with the answer, be it practical or scholarly understanding.

Regardless of what discipline one's research is within, the process of voicing research questions is critically important. Without this step, research will be unfocused, as questions require answers, which require analysis and data for support. The research question leads students back into the scholarly literature to determine what past approaches have addressed similar questions. Framing the question will also help the student have clearer communication with her or his major professor who has more experience reading about and writing research in the field.

Writing About Questions

Topics can be narrowed by looking for opportunities to improve methods or to apply established approaches in new contexts. Method development might include adoption of new variables, use of a different approach, or reassessment of a previously studied topic with additional data. For example, Geographic Information Systems have enabled geographers to approach old and new data in novel ways by integrating multiple forms of information.

1 *The Craft of Research* (Booth et al. 2008) is an excellent resource for further reading on topics and questions.

Similarly, addressing a topic in a novel location incorporates a different context, leading to comparable results to those of other studies in other contexts. Economic patterns that drive real estate markets in one city, for instance, might lead to ideas for how to study the same topic in a different city, providing a basis for comparison. Method development and studying new contexts help the researcher identify gaps in knowledge, but realizing and communicating the importance of the research requires developing questions about the topic.

An example of one of my research questions is; "did animal species selected by hunter-gatherers living in western Argentina during the last several thousand years change over time?" A follow-on question is, "If so, why did species selection change?" The answer to the first question is either yes or no, and I was able to establish that previous studies documented a change in species selection. The second question requires further analysis because perhaps the environment changed over time supporting different types of animal species, thus leading to a shift in what was available for hunters to exploit. Or, perhaps people hunted one or another species intensively, causing its abundance to decline (Figure 3). Alternatively, perhaps people's preferences changed from one type of meat to another. People may have split into groups with protected territories over time, requiring that some groups hunt in new areas or in smaller portions of territory changing what was available to them. Writing the topic into a question led to a flood of potential answers, which became directions for the research.

Answering a research question requires data and analysis to produce evidence. A potential answer may or may not be supported by the evidence. The research objectives for a project, thus, flow directly from knowing what evidence should be mustered to determine whether a particular answer can be supported or discarded. The Question Identification Exercise provides a guide for reviewing a case study within one's field for practice in recognizing and restating research questions that other researchers have studied. The Question Essay Exercise enables new researchers to practice framing questions related to their topic and to develop objectives to move forward.

The Question Identification Exercise

This exercise follows a general format laid out by Booth et al. in *The Craft of Research*. Choose a case study on a topic that relates closely to your own research. This should not be a review paper or a theoretical paper; for this exercise it is better that the paper is a research report that addresses one or two important questions. Profile the study by answering several questions. First, what journal is the paper published in and what is the mission of that journal? Answering this question can help you determine what audience the author is trying to reach. What are the author's key words listed at the beginning of the paper? What are three to five additional action words that you can use to describe the research? Investigates, examines, analyzes, concludes, reflects upon... are examples of action words (see Booth et al. 2008:39 on action words); use

Figure 3. Dr. Miguel Giardina, a colleague from Argentina, who captured a small armadillo (piche) by hand. This species was commonly exploited by hunter-gatherers in the past and is hunted by local smallholder ranchers (puesteros) in the region today.

them to describe what the author does in the case study. Describe the research by combining the action words with the key words. Perhaps the paper is on changes in soil nutrients in a region related to fluctuations in climate; key words might include soil nutrients, precipitation, leaching, and climate change. A description of the topic then is "this paper investigates changes in leaching of soil nutrients due to shifts in precipitation related to climate change."

Next, summarize the major theme or event covered in the paper; for the above example this would be "environmental impacts of climate change." Then, note the study area in which the research took place. If the research took place on the Gulf Coast of Texas, this can be added to the topic description and the theme. Your next step is to triangulate from the topic and theme to the research question(s); do this by adding a phrase to the end of the topic description that states, "because the author wanted to find out what occurred, how something occurred, or why something occurred." The case study might have been warranted using any of those reasons or all three. The topic-sentence example in the previous paragraph would become "this paper investigates changes in leaching of soil nutrients due to shifts in precipitation related to climate change because the authors wanted to find out if global warming is causing leaching and depletion of soil nutrients near the Gulf Coast of Texas."

Now, state the significance[2] of the topic by adding a statement that communicates, "to help the reader understand..." For example, write, "The research was done to provide the reader with a geographic understanding of the long term impacts of climate change on soils in the region." You can now easily rephrase the topic sentence as a research question, "What are the long term effects of changes in precipitation on leaching of soil nutrients along the Gulf Coast in south Texas?" As Booth et al. (2008:57) state, there is a particular value to voicing research as questions, "they force you to state what you don't know or understand but want to." What you want to understand is termed "the research problem," which can be applied or pure. Stating the research question, thus, is particularly important because it helps you describe the consequences of not addressing the research problem or not answering the research question. The *consequences* are the practical or scholarly costs of not doing the research (Booth et al. 2008:53–61). Once you have voiced the research question, state the consequences of not answering it. In this example, the consequences could be far reaching; if we do not understand the impacts of changes in precipitation on soil leaching, we might not be aware of long term changes in soil fertility, which could have further environmental and economic consequences (e.g., reduction in crop productivity or carbon sequestration). Identifying the consequences of not answering a research question and practicing with published case studies enables you to envision gaps in knowledge; voicing research in terms of consequences is particularly beneficial when explaining the value of one's own research. The Question Essay Exercise puts the skills learned in the Question Identification Exercise to work in one's own research.

2 Assessing the significance of a research question is covered by Booth et al. (2008, Chapter 4); I use the prompts that they developed on page 51 of their book in this section.

The Question Essay Exercise

In this essay, you should be narrowing down precisely what your research will cover (Appendix 1, Exercise 7). Start by writing an introduction to your question(s). Your introduction should briefly contextualize each question into the topic, research area, and literature on similar, previous research. Your essay should also communicate the consequences of not answering each research question. Once you have identified questions, move forward with a self-assessment. You are beginning the research proposal stage of your graduate program; your thesis proposal will introduce readers to the field, to your topic and related ones, and then zoom into your research questions. In the proposal, you will be able to utilize voicing of consequences to communicate the need for your project, which relates to the gap in knowledge that the topic fills.

You should be able to frame research questions by early to mid-second semester in your graduate program. Your previous work on the annotated bibliography, summary of interests, and topic selection, along with guidance from your major professor, will have helped you identify important theoretical precepts and methodological approaches in the field. Your self-assessment should characterize the tasks necessary to learn concepts and methods to make progress. Set target dates for when you hope to have skills and knowledge mastered for particular tasks. For example, you may need to learn a particular statistical approach that is commonly employed in your research area before being able to analyze data. Alternatively, you may need to learn one or another fieldwork method. Describe what these approaches are, provide an indication of what will be required to learn them, and set a target date for mastery. All tasks that relate to pursuing the research should be itemized similarly. Meet with your major professor and ask him or her to review your plan to see if anything is missing and to gauge the appropriateness of the outlined tasks and timelines.

Talking Points

As you move forward in your graduate program, it will become increasingly important for you to be able to communicate about what you do and why it is significant. Booth et al. (2008:34) refer to this as developing your "elevator story." They mean that you should know how to summarize and discuss the importance of your topic so that you could discuss it meaningfully during a conversation that lasts about as long as an elevator ride. Like writing, this too represents experience with communicating about your research, and it reinforces motivation. When a researcher is able to provide a summary of their topic in a way that engages another person, the result is often that they feel supported in their endeavors. When research is tucked away in a laboratory or office, behind a computer, a researcher's enthusiasm for their topic more easily wanes. As is clear in the transition from annotating sources, to refining interests, to trimming down to a topic and questions, one's research seems increasingly narrow. A natural response to this narrowing process is that an early career researcher may think their research is too limited to retain its broader implications. However, to gain control over a subject matter that enables an ability to become part of a learned community of experienced researchers, students must first engage in the uncomfortable process of framing their own research questions. As Booth et al. (2008:41, emphasis in original) put it, "If a writer asks no specific *question* worth asking, he [or she] can offer no specific *answer* worth supporting." How does one get to a point at which she or he can frame their research question in a way that others can recognize that the answer is worth supporting?

Undoubtedly, your research will come up in a variety of situations from family and social gatherings to professional meetings. When asked about "what you do," have your elevator story ready. To do this, you have to articulate questions (see exercises earlier in this chapter), but the most important piece of the communication puzzle is envisioning consequences of the research. Practice writing about consequences by focusing on what is left unknown if your questions are unanswered. During conversations about your research, it will be important to have a precise, brief description of your topic, and then statements such as, "if questions related to this topic are left unanswered, we continue to lack understanding of...."

The most important audience for your elevator story in terms of career development is the community of researchers at professional and scholarly conferences. Depending on their interests, your elevator story will only be a starting point in many conversations. It is likely that more experienced researchers will know about the gap in knowledge your topic will fill. In addition, they may have additional research questions in mind and ideas or comments that can support your project. As you move forward in your career, this community will become increasingly important. If you have no questions worth answering, which is actually quite common for students, the researcher you are talking with will notice and will have a bad first impression, and as the saying goes, "you only get one chance to make one..."

Becoming a member of the research community in your field is *the critical transition during graduate school.* As a result, developing talking points is simply good practice for becoming conversant about the consequences and importance of your research. Develop your talking points, frame your elevator story, and never pass up an opportunity to talk about what you do (see Appendix 1, Exercise 8).

Framing Research Objectives Exercise

Knowing the consequences of your research has an additional benefit. You will next consider what will be required to provide an answer. Such consideration represents framing research objectives, and to do so you must consider, "what must be done in the research?" "What data will be required?" "What methods will obtain those data?" "What theoretical concepts are important for articulating the research?" To answer these questions, look back at what other researchers have done in studies on your topic. By now, engaging in your reading and writing routine should have built substantial awareness of ways that researchers tackle such problems. In this exercise do two things: 1) itemize data requirements, theoretical concepts, and methods required to do the research; and 2) begin constructing narrative (e.g., sentences and paragraphs) that summarizes how to access data, how theoretical concepts frame the research, why methods are appropriate, and how to employ them. Congratulations! You are now well on your way to planning your thesis.

Summary

This chapter is all about the research question. More precisely, it is about using the consequences of not answering your research questions to communicate their importance. Your topic represents a gap in knowledge; practice identifying and describing questions that provide answers that fill such gaps by reading and summarizing those aspects of published case studies. What new understanding stems from the research? Once you have worked with the Question Identification Exercise, turn your focus to your thesis research. The Question Essay Exercise takes you through the same process related to your own research. When asked about your thesis research, be able to extract a number of talking points from the Question Essay. Condense these into an elevator story, which will enable you to practice communicating about your thesis with many audiences. Finally, the process of focusing on a question and consequences makes it easier to articulate research objectives. At this stage, you should be outlining your thesis proposal and making plans to do the research.

Chapter 6: The Literature Review

Narrowing to the topic, questions, and consequences of research followed by framing objectives is a significant portion of the journey to becoming a researcher. The reading and writing routine, when established in the context of that narrowing process, represents a progressively growing corpus of experience in a field. As a student turns to her or his thesis proposal, she or he should be well on the way to being a member of the research community. As a result, the student should be ready to compose a literature review related to his or her topic.

There are many perspectives about what a literature review entails. Many researchers hold that such a review should be lengthy and exhaustive; I do not share that perspective. The literature review should provide the context for the topic, questions, and objectives. That is, it should be tailored so that it presents the gap in knowledge that the research will fill. In a sense, the literature review makes a claim, stating why the proposed research should be undertaken. It also addresses how the research should be done. Once the review has provided this context, its purpose has been met. Rather than writing an exhaustive review, the author should compose one that is strategic and that sets the tone for the importance of the research. The literature review should follow an introduction to the research topic, question, and objectives at the beginning of the proposal; thus, it is the second major section. I divide the literature review into four sections: indicative case studies, the theoretical framework, important methods, as well the merits and impacts of the research.

Indicative Case Studies

The intellectual merits and broader impacts sections of the literature review focus directly on the gap in knowledge that the research fills. This cannot be done well unless the researcher prefaces those sections with examples of research that highlight the gap. Summarizing important case studies on the research topic shows what has and has not been done. The Interest Essay confronted a similar challenge and represents practice for this section of the literature review. Summarize roughly five important cases studies to showcase previous research on the topic. Each summary will pose the question asked in the study, the consequences of the research, the method employed, and the conclusion that was drawn. The summary should also describe the context of the research, such as where it took place and its theoretical perspective. Once summarized, it is important to state directly how the study advanced the field.

In your selection of case studies, be certain to cover a variety of approaches, perspectives, and contexts. If you are unable to provide adequate coverage of the topic, provide additional summaries. Your research will fill a gap in knowledge between the contributions of other studies in the field. Use this section to show the spaces left by previous research. Conclude by clearly stating the gap in knowledge that your research will fill, which helps the reader relate the literature review to the objectives of your research. Finish this section with a transition sentence that leads into the summary of theoretical frameworks commonly employed in the field.

Theoretical Frameworks

Some areas of research are mainly (or purely) methodological; however, many researchers employ a theoretically prescribed set of principles to frame their research. Areas of ecology, for example, have detailed theoretical principles or working assumptions that must be clear before pursuing research. The same is true for many of the social sciences (e.g., anthropology) and behavioral sciences (e.g., economics and psychology). Often these frameworks rely on central concepts that support a particular paradigm in the field. Community ecologists, for example, may or may not employ the concept of "succession." Cultural anthropologists may adopt a theoretical perspective that culture is a kind of "text," which ethnographers must interpret. They interpret such texts as relations of discourse concerning the dynamics of power in a society. Alternatively, and often diametrically, some cultural anthropologists adopt an explicitly ecological perspective by framing theoretical concepts such as the adaptive aspects of culture. Regardless of what field one finds a home in, it is important to understand whether there is a variety of theoretical foundations informing research. It is also important to choose one of those frameworks and to be clear about its principles and assumptions during research. The researcher must provide clear definitions of concepts used within that particular paradigm. Indeed, methods may also vary among theoretical frameworks.

Summarizing Important Methods

The first two sections of the literature review (Indicative Case Studies and Theoretical Frameworks) provide the focus for choosing research methods. Within economic geography, for example, there is a diverse set of theoretical perspectives. A critical geographer who studies the impacts of neoliberalism on the developing world is more likely to adopt a Marxist, culture as text, or other post-modern theoretical framework for their research, which maintain that the impacts of economies are embedded in the subjective narratives of people's lives. As a result, critical geographers employ ethnographic methods, such as participant observation and thick description (the detailed recording of one's own experience when living with and learning about a community). If the researcher is interested in the impacts of neoliberal economics on rural farmers in an area (e.g., communities in Nepal, Ghana, Peru, or another part of the world), he or she will live in that community for an extended period of time to learn as much as possible about the community. The result will be an account written by the researcher in which she or he seeks to translate the text of individuals in the community of interest.

A business geographer, in contrast to the critical geographer, adopts a considerably distinctive approach to economic geography. Business geography employs the principles of location allocation and the tools of GIS to help businesses determine where to locate parts of their operations. This includes the best places to place retail outlets, distribution centers, or headquarters, as well as optimization of distribution and/or delivery routes. A business geographer might be interested in the impacts of social, demographic, and economic change on one or more sectors of the economy. Due to the principles adopted and the ways that data are collected, business geography (and many other types of economic geography) employ a variety of quantitative analytical approaches, which include elementary and multivariate statistics. The critical geographer would view a quantitative approach as too objective for the types of research questions they ask, and a business geographer might find that accounts from ethnographic research do not fully support their claims about location and logistics.

One cannot choose a theoretical perspective that does not come packaged with related methodological approaches; thus, to some degree, the choice of theory constrains methods. The summary of indicative case studies earlier in the literature review provides a powerful way to make and defend selection of theory and methods. In addition, rely on your major professor for advice regarding understanding and articulating the merits of the theoretical framework you adopt. Further, your major professor has more experience using various methods; seek advice on which methods to use and how to write about them.

Merits and Impacts of the Research

Summarize the preceding sections on indicative case studies, theoretical frameworks, and important methods. Use that summary to articulate the context of your research. "Merits" refers to what your research will bring to the research field. Are you applying a theoretical approach to a place or within a topic in a way that has never been done? Are you developing a new methodological approach to consider a research problem in a new way? Is your research taking place within a never-investigated context using established theoretical perspectives and mainstream methods within the field? What is it that makes your research valuable? The Interest, Topic, and Question essays provide a means to navigate toward a more focused research plan. The role of the literature review is to highlight how your research fits into the field by articulating the gap in knowledge it will fill; it represents a summary of previous research on the topic against which you can cast the merits of your research.

The literature review will also help you articulate the impacts of your research, which are benefits that extend beyond the field. A way to recognize impacts is to ask and answer,

"How will my research benefit society?" A pitfall with impacts is that some researchers are rather Pollyannaish, meaning that the impacts they summarize are so general that they are essentially meaningless. Summarizing the impacts of critical geographic research on neoliberalism and local farmers in developing countries as "countering pervasive historical impacts of colonialism" is an impact, but does it articulate the precise benefits of the research? It is so general, that it dilutes the perceived impact of the research. Instead, such research might help international non-profit organizations recognize one or more local communities needing aid. In addition, narrative accounts from one or more communities might help aid organizations determine precisely what is needed in those contexts. It is important to pinpoint realistic impacts.

As with the preceding sections of the review on previous studies, theory, and methods, stating the merits and impacts of the research helps the reader understand the context of your project and its importance. As with any other aspect of research writing, it is important to maintain an open dialogue with your mentor as you consider merits and impacts.

Literature Review Exercise

This exercise should help you assemble the components of the literature review. First, settle on a list of 15 to 20 sources that are important case studies in your field. Rely on your annotated bibliography for this step. Re-read those sources, and narrow the list down to ten sources that are the most important ones from the previous set. Then, set a meeting with your major professor and discuss with her or him the original and shortened lists of case studies you are viewing as indicative case studies that relate to your research question and project objectives. Communicate to your major professor that you are following the format provided in this chapter for your literature review for a couple of reasons: 1) your mentor may want you to format the literature review differently, but the components described in this chapter will still be helpful. 2) Your mentor may consider other sources that provide important applications of theory or methods that should be included in your section on indicative case studies. It is better to get your major professor's input on these two points early in the process so that you do not have to make substantial changes later. Recall that what is presented in this chapter is only one of many potential ways to frame a literature review.

Next, describe the theoretical framework you have chosen. Write a summary of the important tenets of that framework, and define important concepts. Make a list of important sources that relate to the theory. Then, make a list of the important methods that you will be using to address the research objectives of your project. Write two to three sentence summaries of what those methods are and how they have been employed in similar research. Create a list of important sources that have used the method(s) to pursue similar research. Then, set a second meeting with your major professor to receive her or his input on your summaries of the theoretical framework and methods. Make sure you send your summaries ahead of time, and ask her or him to consider whether you have missed any important concepts, approaches, or sources.

Finally, write a short two to three sentence description of the gap in knowledge that your research fills, stating how your literature review reveals and contextualizes that gap. Use your articulation of the consequences of your research questions from previous exercises to highlight the merit of your research. Then consider whether there are benefits that extend beyond the field, and write about those impacts. Compare these statements to your elevator story, and see if the story needs revision. Set a third, brief meeting with your major professor to share the merits and impacts of your research; send your summaries ahead of the meeting and ask her or him if they see additional merits and impacts that you have not thought of. When you have finished, you should be able to synthesize these tasks and write a more thorough literature review for your thesis proposal.

Summary

The focus of this chapter is an important component of the thesis proposal, the literature review. Too often, students write bland, exhaustive, and directionless summaries of the literature in their field. The problem with such a literature review is that it is not oriented toward a purpose. From the reader's perspective, such a review may be minimally helpful as it summarizes sources from the field; however, an important opportunity is lost when the review detracts from the rest of the proposal. If the graduate researcher orients the review toward engaging members of the audience with the context for research questions and objectives, the literature review supports the rest of the proposal. Deconstruct the literature review into four components, which can be used strategically to contextualize the research. First, narrow down important sources to a limited number of indicative case studies that help identify the gap in knowledge that the research will fill. From those sources and through conversation with the major professor, identify and choose a theoretical framework, which can be introduced in the literature review. Frame important methods in the same manner. Finally, articulate the merits and impacts of the research by recognizing the gap in knowledge as well as the consequences of the research. The process of writing a literature review represents an opportunity to show the reader the importance of the proposed research in the context of previous studies. It also represents a significant opportunity to interface with the major professor and to capitalize on previous exercises from the annotated bibliography to the various essay exercises presented earlier in this book. Once a draft of the literature review is written, the researcher has constructed one major portion of the thesis proposal.

Chapter 7: The Proposal Storyboard

A thesis proposal should be designed before it is written. Researchers frame what they propose in particular ways that engage an intended audience, one that needs to be convinced that the research is worth doing. For the graduate student, the audience is the thesis committee. What is to be proposed is a project that demonstrates an ability to become an independent researcher. The proposal must communicate the basics of the field that a student is interested in, the types of research topics and questions that are common in the field, as well as a gap in knowledge that the student hopes to fill with their research, but it will go farther than setting up the context of the research by organizing action. It is helpful to conceptualize the proposal as a whole comprising integrated parts (see exercises in Appendix 3). Each part has a specific purpose for communicating about aspects of the thesis research. Briefly, the proposal will have the following components: an introduction, a literature review, a section on research objectives, a description of research methods, and a presentation of a research plan. The proposal should end with a summary statement that reiterates the thesis objectives and the research plan. These sections are discussed in more detail below.

It is very difficult to simply sit down and write a research proposal from beginning to end. Instead, envision the various components, and outline them. An outline provides the skeleton of the sections of the thesis proposal; it can easily grow into a storyboard, however. A storyboard is an expanded outline that leaves space for notes, lists of tasks to be accomplished to finish a section of the proposal, and eventually even narrative that is added for a section as it is drafted (see Booth et al. 2008:130–131 for a separate description). Over time a storyboard should morph into the proposal draft as tasks are completed. Use of a storyboard allows the researcher to see which parts require development and which ones are close to completion. As a result, the storyboard functions not just as a means to organize one's thoughts but also as a way to monitor progress. As with the reading and writing routine (Figure 2), monitoring progress leads to incremental instances of satisfaction that relate to completion of smaller tasks. This helps the student stay motivated because rather than having the goal of "writing the proposal," the researcher can focus on a task, such as, "draft the section on consequences of the research question in the proposal introduction." Use of the storyboard helps the student remain aware that the thesis proposal is an organic whole but also to see that it is the sum of many parts. If the researchers did each of the exercises presented earlier in this book, she or he will see that the foundations for the thesis proposal are already prepared.

The remainder of this chapter breaks the thesis proposal storyboard down by section, which provides a framework for itemizing tasks. The storyboard approach represents a process in which the student learns to impose structure on the unstructured challenge of independent research.

The Proposal Introduction

You will eventually write a very fluid introduction to your thesis that finds common ground with readers. For the proposal, however, it is important to begin the introduction with a direct statement about which field or research area that the thesis will fit within. Answer, "What is the research topic?" That is, start with the gap in knowledge that you plan to fill because you can then provide a sense of direction for the reader within that topic; you can also easily link the topic to a gap in knowledge, which will help convey why the topic is important. Make sure this part of the introduction describes what the research focuses on, where it will take place, and briefly how it will be addressed. Revisit the Topic Essay Exercise from Chapter 3 for a refresher on how to communicate about your topic.

The second task within the introduction is to communicate the research questions that will be addressed within the topic. There may be one or a few important questions that your research attempts to answer; here, I write as if there is only one question. Revisit the Question Identification and Question Essay exercises from Chapter 5. Briefly, articulate questions about the topic by considering what needs to be done to fill the gap in knowledge; answer, what needs to be understood if research on the topic is to be completed? Recall that voicing questions about a topic provides a means to think about having an answer. To answer a question, one must meet objectives. The question relates directly to the negative consequences of leaving it unanswered. If, for instance, I am working on the topic of "improvement of methods for obtaining and identifying protein food residues from ancient cooking pottery," I can voice several questions: "Does one type of method of ancient protein analysis work better than another one?" Or, "Does the ability to retrieve proteins from cooking pottery differ depending on how long the pottery was buried or with the conditions of burial?" (Figure 4). Many other questions can be asked about this topic. The point is that asking one or another question enables the researcher to think about what will be required to provide a convincing answer. It does not matter if the topic is ancient protein analysis or the impacts of big box stores on central business districts in small towns in Texas; ask multiple questions that require answers to gain focus.

Figure 4. Experimental pottery used for method development in our laboratory. Photo by Jonathan Dombrosky.

Turn the question toward its significance. Your thoughts should be, "I am asking this question so I can understand..." (see more on this strategy in Booth et al. 2008:45–48). In the example from the previous paragraph, the significance can be stated as, "I am addressing research questions on ancient protein residues from cooking pottery to better understand whether or not such residues preserve well and can be used to understand the diets of ancient people." Stating the significance of answering the question conveys the research as a problem, which presents a condition that if left unmet leaves the question unanswered.

Next, convey the question/problem in terms of the consequences of the research not being done; "if this question is not answered, members of the research community may continue to ignore this important form of data about past human diet." From the consequences, one can lead directly into stating the objectives of the thesis research. The objectives are the requirements for answering the research questions, which will circumvent the negative consequences of leaving the question unanswered.

In summary, the proposal introduction is a series of telescoping writing tasks that narrow from the topic down to the question and consequences to the study objectives. The introduction section of the storyboard should have the following tasks:

1. Describe the topic.
2. Write about the topic voiced as one or more research questions.
3. Write about the significance of the research question.
4. Write about the consequences of not answering the research question(s).
5. Write about the study objectives needed to answer the question(s).

The introduction should conclude with a brief road map, which tells the reader what to expect, section by section, in the rest of the proposal. It will be something such as, "to meet the study objectives, there are several methods that must be employed, which is followed by a detailed research plan; however, first it is important to contextualize the proposed research into the field in a literature review." Not only does this provide a road map, but it also provides a transition to the literature review by stating its purpose.

The Literature Review

The introduction sets up the proposal, but the literature review contextualizes the topic, research questions, and study objectives into the field. Details of how to design and construct a literature review are presented in Chapter 6, but it is important to learn the critical role that it plays in the proposal. There are two important roles played by the literature review. First, in contextualizing the research the review communicates baseline expectations about what research is like in a particular field. This is important for readers from outside of that field who may adopt distinctive theoretical frameworks and methods. To use a metaphor, the literature review sets up the rules of the game and dimensions of the field of play that the proposed research will take place within. The graduate researcher has learned these rules from extensive reading of and writing about the literature.

Second, the main audience of the proposal is the thesis committee. Members of the committee need to feel confident that the graduate researcher is well on his or her way to becoming versed in the norms of the field. The transition from student to researcher should be almost over by the time of the proposal defense, and the committee members are hoping to see expertise in the field, particularly related to the thesis topic and the gap in knowledge it will fill. The literature review provides this evidence. There should be a fluid transition from the literature review into the following section on research objectives.

Study Objectives

The proposal audience has just read the literature review and has learned the context of the research. It is important to redirect the reader toward the topic and research questions leading into stating the research objectives. Therefore, start the research objectives section by briefly restating those and also the consequences of not answering the research questions. It is important to think of "answering the research questions" as a set of research problems. That is, articulate what problems need to be solved to provide meaningful answers to the research questions. Objectives are clear statements that communicate what needs to be done to solve those problems. For the ancient protein example described earlier in this chapter, there are several objectives. Recall the topic, questions, significance, and consequences:

Topic: "Improvement of methods for obtaining and identifying protein food residues from ancient cooking pottery."

Questions: 1) "Does one type of method of ancient protein analysis work better than another one?" 2) "Does the ability to retrieve proteins from cooking pottery differ depending on how long the pottery was buried or with the conditions of burial?"

Significance: "The research is important for addressing research questions about ancient protein residues from cooking pottery to better understand whether or not such residues preserve well and can be used to learn about the diets of ancient people."

Consequences: "If these questions are not answered, members of the research community may continue to ignore this important form of data about past human diet."

Study objectives include:

Objective 1: Evaluate potential methods for ancient protein analysis from previous studies.

Objective 2: Create experimental cooking pottery that can be buried in distinctive soils for different lengths of time (Figure 4).

Objective 3: Apply the various methods from Objective 1 to experimental pottery from Objective 2 to determine which ones are optimal for studying ancient proteins.

Note that objectives guide the research; they represent goals, not plans. There will be numerous tasks required to accomplish the study objectives. Objectives are important because they provide a link between the gap in knowledge framed in the topic, questions, and consequences in addition to a research plan of action items (tasks). Thus, the objectives provide a focus for the research; without them, many students pursue tangential, seemingly related tasks that may not answer the research questions. Framing objectives requires a commitment to act in a particular direction.

Proposed Research Methods

At the proposal stage, methods represent the parameters of actions that will be taken to accomplish the research objectives. The methods section will include sub-sections on data requirements, data collection, and analysis. Depending on the type of research, there might also be a section on the study area.

Study area—in many areas of geography, anthropology, and other social sciences as well as ecology a major component of the context for a research project will be the place or community in which it will take place. Indeed, the study area may be the reason for the gap in knowledge, as it could be the place or community that has never witnessed previous research on a topic. Description of the study area will include all aspects of the place or community that relate directly to the research. If the topic falls within the economic geography of a particular type of business in a community, then the economic, political, and social aspects of that place that are relevant to that type of business need to be described. If the research is on a topic that relates to environmental biology or geography in a place, then local data on weather, soils, topography, vegetation, and animal communities may be required to describe the study area. It is important to think about what the audience for the research will need to know about a place or community when describing the study area.

Data requirements, collection, and organization—this section provides the audience with a description of the data that will satisfy the study objectives. For some types of research, this may require an entire section on experimental design. For example, for the ancient protein research, there would need to be presentation of the tasks required to implement cooking and burial experiments as well as laboratory processes for studying food residues. In contrast, when a student is conducting fieldwork, all phases of that work will need to be itemized as tasks. For example, a research project that focuses on studying biodiversity of freshwater mussels in streams upstream from, within, and downstream from an urban area, would require a field sampling design, incorporating when, where, and how often to sample. Alternatively, some studies, particularly in fields such as medical geography, rely on data mining from existing large datasets (e.g., census data). In such studies, the researcher would provide metadata on the source, the details of important variables that will be studied, and how the database will be queried. The point of writing about data requirements is to demonstrate to the reader that there is a strategy for how data will be collected to meet the study objectives. Here are some steps to communicating about data requirements:

1. Communicate the nature of the study; for example, is the research field, lab, or data-mining intensive?
2. Define multiple variables that must be studied to provide answers to the research questions.
3. Identify data sources by describing precisely how, where, and when data will be acquired.
4. Describe the experimental design, the field sampling design, or the data-mining approach.
5. Discuss which types of data will be used for particular types of analysis. That is, state how each type of data will be used, which provides a transition to describing analytical methods.

Analytical Methods—the researcher will not be able to satisfy the research objectives once data have been produced from the study's experiments, fieldwork, or data mining unless she or he chooses an analytical strategy. The literature review will have summarized common analytical approaches in the field related to the topic. In addition, the major professor will be an expert in one form of analysis or another. An important part of graduate studies is to learn such analytical approaches so that they can be applied in novel ways to research questions in the thesis. A student who designs an ethnographic field study for example, will learn qualitative approaches, such as discourse analysis to interpret culture as text drawn from interviews with members of the study community. Experimental laboratory research in, say, environmental chemistry will use a standardized toolkit of statistical tests for interpreting the meaning of results. A project that focuses on remote sensing of landscapes to study one topic or another will draw from a variety of spatial statistical tests. For the early career researcher, the most important thing to remember is that the choice of analytical approach is one to make carefully. A common pitfall is to use too many types of methods or statistical tests, thinking that one cannot over-interpret data. Experts in the field, however, know that some approaches are better than others for answering one type of question or another; if you fail to make a choice (or if you make the wrong one), experts will know it. Here are some tips:

1. What type of data is being used in the study? Qualitative? Quantitative? Categorical? Ordinal? Multivariate? If you do not know, it will be important to brush up on qualitative and/or quantitative methods.
2. What are common approaches and/or tests that are used by researchers conducting similar studies?
    a. Do you have similar data to those from other studies on the topic?
3. What analytical approach does your major professor recommend?
    a. Have you learned it?
    b. Ask your major professor for additional case studies that use the approach.
    c. Find alternative datasets to work with to practice using the approach.

Writing about research methods helps provide a focus upon tasks for completing the research. Such writing represents "knowing what to do;" it does not, however, constitute an action plan that breaks down the proposal into a series of tasks. The student should round out the proposal with a research plan that breaks down each component of the thesis into manageable tasks.

The Research Plan

I have read many proposals and attended a number of proposal presentations that end with a short summary of the proposed research after a brief methods section. The message seems to be, "it is obvious what needs to be done." The student has missed an important opportunity to accomplish two things. First, they could be setting a more precise plan with tasks that is easier to monitor. As with the reading and writing routine for the annotated bibliography and essay exercises, the student could establish a thesis routine. Second, providing a plan with tasks and a timeline demonstrates to the thesis committee that you are ready to move forward (beyond what seems obvious). That is, they will be impressed with a thesis plan because they see them so rarely.

The best way to establish a thesis plan is to deconstruct each portion of the storyboard, breaking it into tasks. Much of this has already been done in this chapter with numbered bullet points. However, this can be refined by determining subtasks. It may be, for example, that the student knows which analytical approaches are required but has not yet mastered the method. Tasks related to the Analytical Methods portion of the research plan could include "read case studies that employ the method," or "practice with alternative datasets to learn method." Each stage of the process needs to be broken down similarly, outlining tasks that relate to learning, data acquisition, analysis, interpretation of results, and thesis writing. A detailed plan is better, one that considers how much can be accomplished weekly and monthly as one moves from the proposal toward thesis completion.

It is critical to set target dates for completion of tasks. The plan should provide general dates that anticipate completion of major milestones, such as "completion of fieldwork" or "learn and practice analytical methods" or "complete data collection" in the proposal defense, but the plan should be much more detailed and revised weekly thereafter. It is important to be flexible when revising the plan; there will be target dates that are not met. Simply set new ones to keep making progress (see Destigmatizing Deadlines in Chapter 4). The point is to envision manageable tasks and to recognize small milestones. Monitoring helps the researcher stay motivated; so keep track of what is accomplished each day, each week, and each month. This organized approach will lead to a timely completion of the thesis.

Summary

The proposal storyboard is more than an outline. It is a living document that breaks down the proposal by section. Each section will change over time, starting with writing tasks and ending with a full draft of the proposal. Once written, sections should be inserted into the storyboard, which enables the researcher to monitor progress toward proposal completion. Incomplete sections of the proposal should have lists of completion tasks in the storyboard, such that it is easy to focus upon what needs to be finished. Students should draw upon the work they did when writing essays for exercises from previous chapters to complete their storyboard. Once you complete the proposal draft, send it to your major professor for comments. Once she or he is satisfied with your revisions, you will be able to schedule your proposal defense.

Chapter 8: The Proposal Defense

Scheduling the thesis proposal defense varies by program and major professor; a common approach is for the major professor to review multiple drafts of the proposal. After several rounds of revision, a "defense draft" will be produced that is circulated to the committee. Depending on the major professor, the student might wait for feedback from the committee before setting the proposal defense date. Alternatively, some programs encourage the student to set the public defense date once the thesis committee has received the defense draft, holding that any revisions required by committee members will be made after the defense. Regardless of the process, at the time that the written proposal is circulated, it is time to develop and practice the public presentation for the defense.

Students overlook two important aspects of the proposal presentation. First, the audience is different than for the written proposal. Second, the medium for communication is entirely different. These differences are seemingly easy to recognize, but students make a number of mistakes when translating the ideas fleshed out in the written proposal to the defense presentation.

The Defense Audience

The proposal-defense presentation is given to a general audience of students and faculty members in one's graduate program, whereas the audience for the written proposal is the thesis committee, which is a subset of a community of experts in the field. The general audience at the defense presentation, moreover, will include non-experts—ones who are interested and intelligent but who are not necessarily surrogates for the research community in the field.

The proposal defense presentation is analogous to teaching the audience about one's research rather than about engaging members of the research community. The main points of the research must be summarized concisely, and some aspects of the proposal will be dramatically reduced in scope. The literature review, for example, takes on a limited role in the presentation. Identifying the gap in knowledge, the topic, questions, and objectives are key elements of the presentation, and providing a research plan is critical to communicating with this general audience. The presentation can be divided into nine to ten sections, each of which can be presented in a single PowerPoint slide (see Appendix 3):

1. Introduction
2. Research context
3. Research question
4. Study objectives

5. Study area (if necessary)
6. Proposed methods
7. Preliminary results (if available)
8. Research plan
9. Proposal summary

Each section (slide) has a purpose; the goal of the presentation is to voice the proposal as a narrative with the PowerPoint as visual support. The student will need to develop detailed notes about each section, and she or he will need to practice delivering the presentation multiple times.

The Medium

Contemporary research audiences expect that presentations will be given using PowerPoint, which is a blessing and a curse. The good thing about PowerPoint (and similar platforms, such as Prezi) is that it is easy to organize a presentation and to use visuals. The curse of PowerPoint has nothing to do with the platform; it has everything to do with abusing that platform. Thus, there are practices to adopt and others to avoid when developing the slides that relate to each section of the proposal presentation.

First, avoid use of text. Give each slide a short title, such as "Introduction" or "Study Objectives." Then use bullet points sparingly. A critical mistake that presenters make is to present extensive text, even sentences, in slides. It is impossible for audience members to read, listen, and take notes at the same time. Worse still is when presenters turn toward the screen and read the text. Reading from text-heavy slides represents a worst-case scenario for audience members[1]. Such a presentation style indicates that the presenter either does not know the material, has not practiced the talk, or both. Use as few words as possible when designing slides; instead, use photos, maps, charts, and other visuals that supplement the narrative for that section of the talk (slides in Appendix 3, Proposal Slides Outline, for example, are too wordy for a presentation). If you are presenting about the study area, for instance, provide a photo of the place, and briefly describe it as part of the narrative.

Second, for your own purposes design a narrative statement that goes with each slide. Use the following steps: 1) state the purpose of the slide; for example, "the purpose of this slide is to share the study area with the audience." You obviously will not say this to the audience, but be explicit in your notes concerning the purpose of each slide. 2) Articulate three to four statements to communicate related to that purpose; for the study area, for instance, country and region, then community, then aspects of the community that relate directly to the re-

1 It is possible that a presenter could vomit from nervousness in front of the audience, but extreme circumstances aside, reading from text-heavy slides is about as bad as it gets.

search topic. 3) Craft one- to five-word bullet point phrases that summarize each of the statements. 4) Choose a single image that provides an opportunity for an anecdotal talking point, one that relates to the purpose of the slide. Finally, 5) take detailed notes on the purpose of the slide and on its content. It is helpful to compose prose about each slide as well. You will not read the prose to the audience; indeed, it will likely be set aside. Composing prose, however, is a great way to organize your thoughts and to articulate ideas. Many presenters (particularly ones who struggle with nervousness) summarize their notes for each slide into flashcards that they keep handy during the presentation.

Third, practicing the presentation is very important (see next section); after you practice giving your talk, go back to your notes. Review the narrative that you have written for each slide; be critical by asking, "Did I communicate the points related to the slide in a way that benefits the audience?" Learn to pay attention to points that feel sticky, ones with which you feel you struggle. If, with practice, you feel that one or another talking point is still unclear, seek advice from your major professor. Tell him or her, "I am trying to communicate X and Y in this slide, but I am not sure that I am making the statement clear enough for the audience." And ask, "Do you have any advice?" Slide design, crafting narrative, and self-review of presentations becomes easier as one gets better at presenting.

Becoming a Better Presenter

It is important to recognize that some people are better presenters than others. We can conceive of a presentation-skill continuum, ranging from talented and fluid to uncomfortable and awkward. The less a presenter knows about a subject, regardless of her or his native talent for presenting, the more likely she or he will be uncomfortable while presenting. There are four antidotes for presentation discomfort. First, a solid written proposal represents one's knowledge and expertise in the field. The better the written proposal, the more confident the researcher will be when presenting. Second, it is important to spend a lot of time designing the PowerPoint slides for the presentation. As discussed in the previous section of this chapter, concise slides that use images instead of text enable the audience members to focus on what the presenter is saying. Third, if you think that you are a good presenter, do not wing it. That is, do not trick yourself into relying on talent; in front of the proposal audience, under the stress of the defense, you might not be as fluid as you expect. The fourth antidote—the one that trumps all of the others (assuming a solid written proposal, of course)—is practice.

It does not matter how much innate skill a researcher has for presenting. It is more important to set goals for presentation communication and to practice in such a way that a person does the best that he or she can. Practicing the proposal presentation is a great way to improve in this regard.

Presentation Practice Exercise

The most important practice sessions are private ones. I practice research presentations many times by simply giving a talk in front of my computer with the office door closed. I give the talk about once per day during the week leading up to the real presentation. That said, practicing a talk in private becomes more effective with experience. When starting out, it is important to create the same type of support network for presentations as for peer-review in writing. An effective way to support development of presentation skills is to create a peer forum for giving talks.

Start with three or four student colleagues who share interests and values. If you feel that a student colleague does not have the same work ethic or may be unhelpful during review of presentations, then do not start the group with that person. A group of two dedicated student colleagues will be more effective than a larger group with lazy or disruptive individuals. Many people are not comfortable volunteering to present, so offer to give your presentation during one of the initial group meetings. Set a regular (e.g., weekly) meeting time, and plan one to two presentations per week. Make sure that presenters get multiple periods to practice their entire presentation before an important event, such as a proposal defense. Provide simple criteria for evaluating the presentation:

1. On content: were the goals of the presentation achieved?
    a. Is the topic clearly articulated?
    b. Could the audience members understand the context for the research?
    c. Were the questions, consequences, and objectives clear?
    d. If needed, was the study area described in detail, yet clearly?
    e. Is it clear what methods will be used and why they are appropriate?
    f. If there are any preliminary results, was their meaning clear?
    g. Does the research plan illustrate a clear path forward?
    h. Was the summary a good reminder of the main points of the presentation?
2. On presentation style: was the presenter easy to understand and engaging?
    a. Were any of the slides too text heavy?
    b. Did any of the slides seem disconnected from what the presenter was saying?
    c. Did the presenter have any ticks?
        i. For example, saying "um" too much, or facing the slides.
    d. Did the presenter read from the slides, and if so, in what sections?
    e. Were images, maps, charts, and other visuals easy to see and clear in meaning?
    f. Did the presenter avoid eye contact?
    g. In general, did the presenter seem to have practiced with the information?

To maximize the benefit of a presentation forum, it is important to adopt a similar perspective as the one discussed in the Peer Review Exercise. Use practice presentations to improve your peer review skills. Reviewing the presentations of others is an excellent opportunity to learn constructive criticism. Criticism of writing is comparatively distant and indirect. The presenter is more vulnerable as feedback is in person and direct. Here are some tips:

1. Seek to be helpful by maintaining your perspective as an audience member.
2. Ask questions in a manner that seeks clarification, even if you think the presenter is wrong about an idea.
3. If you have a serious criticism about the research, save it for a private conversation after the forum.
4. Be vigilant in making sure that you are putting the presenter first (as she or he is the one standing in front of others after a presentation). Too often audience members are arrogantly critical, which is more about that person trying to look smart than it is about a flaw in the presentation.
5. Pay attention to slide design, use of visuals, and clarity of the narrative, and provide the presenter with fair, concise feedback.
   a. Be precise; for example, ask questions like, "In slide 2, I did not understand how your topic fills a gap in knowledge. Could you clarify that?"

The most important tip is to *put the presenter first* when you are reviewing their work as an audience member. The goal of practice is to help the presenter develop their skills and improve their presentation.

At the Defense

The proposal defense will consist of the student's presentation to faculty members and students in the program, followed by an open question and answer period. If the presentation is engaging and the research is interesting, the presenter can expect general questions from the audience. A strategy is to create extra PowerPoint slides with additional examples and other interesting visuals, such as additional maps and charts that might be useful for answering questions. This is a useful strategy because the format of the presentation requires that the presenter not offer too much content during the presentation—a compromise to limit content increases clarity and allows the audience to focus on a few themes.

Once the open question period has ended, the major professor will ask the general audience to leave, and the thesis committee members will ask questions over the details of the research. There are a few strategies for this part of the defense. The first is to expect that committee members will ask questions that are difficult to anticipate; thus, you (as the presenter) may not have solid answers to all of the questions. Committee members are more experienced and will inevitably think of criticisms, ideas, and questions that expand the research. The best approach for handling unanticipated questions is to listen, to write the question down on a notepad, and to consider which aspects of your research enable you to provide an answer. Your research may have led a committee member to think of a completely new line of inquiry, which may or may not be included in your thesis research. More likely than not, once such ideas are voiced, they end up being nothing other than points of interest that are not directly incorporated into the thesis (but they may provide food for thought for future research).

More important is when a committee member identifies a flaw in a part of the study, which could be in theory, method, background research, or analysis. It is important to listen to such criticisms carefully and to take notes for later discussion with the major professor. It is rare that there is a fatal flaw in the research design if the proposed research has been reviewed through an iterative process with the major professor as described in this book. There is always the possibility, however, that something was missed; simply be open to the criticism.

There are times when committee members are overly critical during the proposal defense. He or she may treat a minor problem as a fatal flaw. Adopt the same strategy: listen, take notes, and accept the criticism. Some faculty members test the ability of a student to take criticism constructively during the proposal defense, which is potentially a major component of the exam. It is important to recall that the proposal defense is a pass-fail exam. After the question and answer period, the major professor will ask the student to leave the room, and the committee will have a closed-door discussion of the defense. Any revision requirements will be deliberated, and the committee will decide on whether the student passes the defense. If you designed your research through the careful process described throughout the chapters of this book, you should be in great shape!

After the Defense

I believe that the proposal defense is the most important milestone of the master's career (or early PhD career). The most difficult challenge of becoming a researcher is leaning into the transition from "undergraduate learner" in a structured environment to "self-directed graduate researcher" in an unstructured environment. The first two semesters of graduate school can vary from aimless wandering through general ideas to waiting for direction from a major professor or other faculty members. Alternatively, the student can experience this intense transition by acknowledging that habits from previous experience in college simply will not cut it. Assuming that the graduate student has taken the route described in this book, he or she will have become someone who knows a lot about research design in his or her field during the proposal writing process. This will include intense exposure to the literature, regular practice writing about research, increasing commitment to a topic, targeting of questions and objectives, and the development of a solid research plan. Passing the proposal defense is cause for celebration, and I advise my students to take a couple of weeks off to rest afterward. The research that has been planned is yet to come, and it is important to take a step back, reflect upon what was learned, and to embrace the next steps in the research process.

Summary

This chapter has presented basic information on what to expect during the thesis proposal defense. Very few students realize that the steepest part of the learning curve is research design during the first two semesters of graduate school. The habits that are set then will either haunt or benefit the student for the rest of their graduate career—indeed, for the rest of their research career. It is imperative, thus, that the student see the process through and finish the proposal on a high note. There are a number of ways to ensure that the proposal defense goes well; communication with the major professor throughout the process from establishing a reading and writing routine to creating the proposal storyboard and drafting the proposal is significant. That relationship represents the foundation for preparing to present the thesis proposal to the committee. In addition, the same habits that underlie peer review of written work will support practice of the proposal presentation. It is critically important to consider the audience when designing and practicing the presentation; peer review of the presentation is an essential part of that preparation. The student can expect the thesis committee members to be intrigued by and critical of the research during the proposal defense. In the end, however, if the student has set the foundation and done all that she or he can do to prepare, the proposal defense is likely to represent a significant milestone on the journey to becoming an independent researcher.

Conclusion

Becoming an independent researcher requires a transition in learning style. The student must become someone who can operate without the prompts and structures of a managed workplace or undergraduate education. In my experience, many graduate students arrive to their program unprepared to make this transition and are shocked when they find it is necessary. The climate of many graduate programs exacerbates the problem in a couple of ways; first, the recruiting process assesses and values learning outcomes that reflect habits and abilities that are valued in the structured environments of workplaces and undergraduate education. This is the record that faculty members have to work with when assessing the applicants' portfolios. Second, faculty members overlook the fact that a transition is necessary to learn the habits of independent research by assuming that incoming graduate students will "sink or swim" on their own. The premise of this book has been that students and major professors give a lot away by making this assumption, and that most students can learn to be independent researchers if they directly confront that graduate school is an entirely new challenge.

To make the transition from undergraduate learner or managed employee to independent researcher, the student must develop new habits—ones that support a framework and that develop strategies for setting routines and learning about their field. Undergraduates typically engage in binge learning, arriving at exams and writing term papers at particular points in the semester. In contrast, the successful graduate researcher will make the uncomfortable dive into the cold ocean of the literature in their field by setting a reading and writing routine. At the onset, it seems that there is nothing to write about, but the student must still learn to write about research in their field. The annotated bibliography exercise provides a place to get started; making it routine gets the graduate student off to a solid start to their research career.

The student will quickly find, however, that their research field is an enormous compendium of diverse theoretical positions, methodological frameworks, and contexts for research. A good strategy is to enlist the major professor right away by requesting several articles, book chapters, and books to read that can help the student focus their interests. Early in the process, the student must learn to identify research topics and to articulate how those topics fill gaps in knowledge in the field. Without this ability, the student will not move forward because she or he will not be able to recognize new gaps in knowledge that represent opportunities for research.

As the student becomes a researcher who can address topics in their field, he or she should pay particular attention to time management. It is difficult to support a reading and writing routine if a person has a poor time attitude; short term planning by setting goals on a daily and weekly basis along with monitoring progress are important. Without daily and weekly goals, the milestones of graduate research are too spread out; it is easier to stay motivated when there are more immediate milestones. It is critical, however, to pay close attention to long term planning related to proposal completion and thesis research. Long range planning represents "keeping one's eye on the ball." It is the mixture of short and long range planning as well as monitoring of progress that provides a framework for the unstructured learning environment of independent research. Once a student masters the habits of routine setting and time management it is easier to keep a positive time attitude, and she or he is well on her or his way to becoming an independent researcher.

The student must not stop at routine setting, however. Selecting a research topic that fills a gap in knowledge represents a serious commitment. As the researcher moves forward, she or he must frame the research as distinctive questions, ones that convey the consequences of the research. Such consequences represent what will be lost in understanding if the research is not finished and what will be gained by doing the research. Envisioning consequences paves the path for articulating study objectives, which represent what needs to be done to accomplish the proposed research.

Along the way, it is important to develop a relationship with the research community in one's field. That community represents the audience for the thesis research. The immediate audience of the thesis proposal is the thesis committee, which includes members who are experts in the field. The student, moreover, should lean into the peer review process by contributing to their graduate student community. Finding members of one's student cohort with similar research interests and creating a support network for feedback on research products, such as essays, chapters, and presentations provides an opportunity to learn elements of peer review. Seek to learn how to provide constructive criticism and how to receive any criticism with a constructive attitude. As a student composes his or her thesis proposal, he or she will develop a mentoring relationship with a major professor. Learning the habits of peer review will make it easier to accept and use criticism from the major professor and the members of the thesis committee.

The study objectives will be contextualized into a literature review, which provides the foundation for the proposed research. The literature review summarizes information about important case studies, theoretical frameworks, methods, and analyses that are critical for filling the gap in knowledge addressed by the thesis. It also provides an opportunity to reflect upon the merits and impacts of the proposed research. The literature review will be integrated directly into the thesis proposal, and products from the Interest, Topic, and Question essays will be used to develop a proposal storyboard. This detailed outline will eventually grow into the thesis proposal. It provides a template for delineating tasks to complete each section of the proposal. Addition of sections of narrative to the storyboard serves as a way to monitor completion. As with other exercises presented in this book, the graduate researcher should be meeting with their major professor routinely to flesh out ideas presented in the proposal. The defense draft of the proposal will have undergone multiple rounds of review, editing, and revision by the major professor. The student can anticipate additional feedback from members of the thesis committee.

The first year of graduate school (for master's students; it may take longer for doctoral students) culminates with a proposal defense, which is a public presentation about the merits of the proposed research. The audience for the defense includes students and faculty members from the student's degree program. The presentation requires the use of a different medium, but as with research writing, it is important to strategize with the major professor and to practice delivering the presentation with a group of peers. Once the student has passed the proposal defense, she or he should be prepared to implement a research plan to complete the thesis research.

The first year of graduate school is an opportunity to face the unwieldy transition to becoming a researcher head on. There will be much to learn during the second year when the student dives into his or her research. It is likely that plans will change and that questions will be adjusted to fit realities of the research that simply could not be anticipated while crafting the thesis proposal. The student, however, will have established new learning habits during the development of the proposal if she or he used the exercises presented in this book. By the end of the first year, the student will not yet have learned all of the skills or collected the data required to complete the thesis. Indeed, he or she will continue to learn new skills and ideas at a high rate, but he or she will have become a researcher.

Sources

Baumeister, R. F. 2013. Writing a Literature Review. In *The Portable Mentor: Expert Guide to a Successful Career in Psychology, Second Edition*, edited by M. J. Prinstein, pp. 119–132. Springer, New York.

Booth, W. C., G. G. Colomb, and J. M. Williams. 2008. *The Craft of Research*, 3rd edition. The University of Chicago Press, Chicago.

Britton, B. K., and A. Tesser. 1991. Effects of Time Management Practices on College Grades. *Journal of Educational Psychology* 83:405–410.

Davis, G. B., and C. A. Parker. 1997. *Writing the Doctoral Dissertation: A Systematic Approach*, 2nd edition. Barron's Educational Series, Inc., Hauppauge, New York.

Driver, L. J. 2015. Dynamics of Stream Fish Metacommunities in Response to Drought and Re-connectivity. Doctoral Dissertation, Department of Biology, University of North Texas, Denton, TX.

Driver, L. J., and D. J. Hoeinghaus. 2016. Fish Metacommunity Responses to Experimental Drought Are Determined by Habitat Heterogeneity and Connectivity. *Freshwater Biology* 61:533–548.

Nabhan, G. P. 2013. Ethnobiology for a Diverse World: Autobiology? The Traditional Ecological, Agricultural and Culinary Knowledge of US! *Journal of Ethnobiology* 33:2–6.

Nicholas, K. A., and W. Gordon. 2011. A Quick Guide to Writing a Solid Peer Review. *Bulletin of the Ecological Society of America* 92:376–381.

Quave, C. L., K. Barfield, N. Ross, and K. C. Hall. 2015. The Open Science Network in Ethnobiology: Growing the Influence of Ethnobiology. *Ethnobiology Letters* 6:1–4.

Raff, J. 2013. How to Become Good at Peer Review: A Guide for Young Scientists. Violent Metaphors: Thoughts from the Intersection of Science, Pseudoscience, and Conflict. URL https://violentmetaphors.com/2013/12/13/how-to-become-good-at-peer-review-a-guide-for-young-scientists/. Accessed on August 30, 2017.

Rafter, M. V. 2014. 12 Time Management Tips for Writers. Wordcount: Freelancing in the Digital Age. URL http://michellerafter.com/2014/05/01/best-time-management-tips-for-writers/. Accessed on August 30, 2017.

Shewchuk, J. R. 1997. Giving an Academic Talk. URL https://people.eecs.berkeley.edu/~jrs/speaking.html. Accessed on August 30, 2017.

Silva, P. J. 2007. *How to Write a Lot: A Practical Guide to Productive Academic Writing*. American Psychological Association, Washington, DC.

Appendix 1 – Exercises

There are numerous exercises introduced and discussed in the chapters of the book, at least one per chapter. This appendix compiles those assignments and a few other ones in one supplementary document. Not all of the exercises presented here are discussed in the chapters, and some of those discussed in the chapters are not duplicated here.

1. Pre-semester annotated bibliography
2. Semester annotated bibliography
3. Time awareness
4. Interest essay
5. Topic essay
6. Degree planning
7. Question essay
8. Personal research profile

Here are links to supplemental essays about establishing routines:

Setting the Routine:
https://docs.google.com/document/d/1aFWstdMxNdHExH9LMDzC_V6U8joIN6r8qqjA-Gl5AW-o/edit?usp=sharing

Monitoring the Routine:
https://docs.google.com/document/d/1PRS_LTuguaB4A8eQzDvsPpn-oE4YTuVah4dxwa7-WqI/edit?usp=sharing

1 – Pre-Semester Exercise

My class meets in the fall semester, and students are often eager to get started once admitted to the program. I send them the following assignment at the beginning of summer so that they have a prompt to start establishing healthy reading and writing habits. It also helps them get started with their mentoring relationship with their major professor; I copy their major professors on the email when I send them the assignment. The due date is the first day of class, and this assignment corresponds to material in Chapter 1.

Starting the Annotated Bibliography

Among the several assignments in the coming semester will be to produce an extensive annotated bibliography of journal publications in your field of research. A great place to start is to read and annotate some of the works by your major professor as well as articles by other scholars that they recommend.

Part 1: For this part of the assignment, you will need to annotate 5 articles by your major professor.

Part 2: For this part of the assignment, you will need to approach your major professor and ask them for 10 critical articles that they feel you should read in your field, right away. Annotate these 10 articles using the same approach you used in Part 1.

Part 3: Start a bibliographic trail by selecting 5 articles cited in the papers you read for parts 1 and 2. Read and annotate those papers using the same approach.

You can find information about what is included in an annotated bibliography at this link: http://www.lib.sfu.ca/help/cite-write/citation-style-guides/annotated-bibliography.

Use the APA (American Psychological Association) citation style at this link for your work and follow the sections in Chapter 1.

All work in this class will be submitted through turnitin.com, which is available through the class Blackboard site. Annotated bibliographies are no different than other forms of writing; the writing must be original. Blackboard will be activated by the due date.

To submit your assignment, do the following:

1. Login to learn.unt.edu
2. Navigate to the course page.
3. Navigate to Course Content
4. Click on View/Complete for the Pre-Semester Annotated Bib. assignment.
5. Upload your work and submit.

Due date 8/30/17 at 5:00 pm.

Advice:

1. If you are lost, check with your major professor to see what articles they recommend.
2. Get used to searching for articles through digital engines, such as Web of Knowledge and scholar.google.com.
3. Follow the guidelines at the link above, precisely.

Standards:

1. All work will be turned in as MS Word document files.
2. All work will be 1 inch margins on the page.
3. Work will be in Calibri font, 11 pt size.
4. Use the default spacing 1.08 spacing for line spacing.
5. Do not provide a title page; just put your name, title of the assignment, and the date at the top of your work.

2 – Semester Annotated Bibliography Exercise

An essential quality of an independent researcher is to be well read in your field. This requires establishing the habit of reading and denoting what you have learned from articles and books. Over the course of the semester you will assemble an extensive annotated bibliography in your research area. The process should require you to independently figure out how to search for and obtain important peer-reviewed articles and books in your research area. An excellent place to start is to ask your major professor, "what are the most important papers that should be read immediately in my research area ?" The next step is to read the papers those papers reference. In addition, however, you will need to search the literature to cover it well. In this process, you can expect to "kiss a lot of frogs" because many of the articles and books you encounter will not match your interests. Most of them will have some bearing on your interests. This assignment corresponds to material in Chapter 1.

Your annotated bibliography will follow the same format as the one for the pre-semester exercise. There will be three sections to the annotated bibliography that you will turn in at different points in the semester.

Pre-semester: due first day of class.

Section 1: due Week 3 – 25 additional resources.

Section 2: due Week 6 – 25 additional resources.

Section 3: due Week 9 – 25 additional resources.

I will be grading the citation and annotation according to the criteria at this link:
http://www.lib.sfu.ca/help/cite-write/citation-style-guides/annotated-bibliography

Effective Reading
In order to accomplish this task, you will need to manage it in the midst of your other responsibilities. Thus, this and other assignments are part of your time-management challenge. You will have to read efficiently. Here are some resources that provide advice:

Sink or Skim? http://www.apa.org/gradpsych/2010/11/skim.aspx

Reading: Improve Your Comprehension and Efficiency:
http://gradschool.about.com/cs/reading/a/read.htm

Speed Reading: Learning to Read More Efficiently:
https://www.mindtools.com/speedrd.html

There are many other online resources; dig around and find what works for you.

3 – Time Awareness Exercise

Keeping track of how you spend your time is important for understanding why you are or are not making progress. For this exercise, you will document your time management practices from the first day of school after we meet. You can use a day-planner type table or chart or you can write down start-stop times for different activities. You should create major and minor categories of activities so you can later aggregate the data if you need to. This assignment corresponds to material presented in Chapter 2, and can be used in combination with the Prioritization and Time Monitoring Exercise presented there.

- Teaching - class time, teaching prep, grading, student meetings, course meetings
- Research - literature search, reading and note taking, data collection, analysis, writing
- Other academic - classes, faculty meetings, lab/group meetings, talks
- Personal - meals, e-mailing, gym, Facebook, commuting

After you collect your data, create graphs and/or tables of the time spent in the major categories. Given that you should be working toward writing and doing research every day, what changes can you make to the way you spend your time? What challenges or obstacles do you face in trying to achieve your goals?

4 – Interest Essay Exercise

Prior to offering this assignment, I assign students to three-member peer-review groups. Recall that I have written this assignment for students in a geography program, so wording would need to be adjusted for programs in different fields. This assignment corresponds to subject matter presented in Chapter 3. The exercises here also relate to the Peer Review Exercise presented in Chapter 4.

This essay will provide your reader with an introduction to the area of research of interest to you. This statement should provide a definitive description of the area (e.g., of Medical Geography, Business Geography, Industrial Ecology, Geoarchaeology, or other areas). Physical geography, human geography, and archaeology are too broad, but coastal geomorphology and applied retail geography are too narrow to count as areas of interest in this essay. The statement should answer: why are you interested in a particular area of geography, geology, or archaeology? Further, it should also introduce several research topics within your area of interest that are of current intrigue in the field and state why those topics are important. At the end of the essay, you must choose a topic that you will develop further and state why it is of particular interest to justify your choice. You must also comment on what aspects of peer review from your group did and did not help you.

Page limits: Minimum 2 full pages; maximum 4 full pages.

Reviewing the Interest Essay:

Content Review

1. What is the research area of interest to the author?
2. Is the research area clearly defined; that is, does the author state what it is?
3. Does the author describe what researchers in the area of interest do?
4. Are examples provided to aid in that description?
5. Is the importance of the research area (either as pure or applied research) reflected upon?
6. Are three to four clear topics within the research area introduced?
7. Is it clear that these are targeted within the research area and they are indicative of research in that area?
8. Could any of these topics lead to targeted, specific research questions?
9. Has the author chosen which topic they will develop and then stated why?

Readership

1. Are key concepts defined?
2. Does the author cite sources that increase your confidence in their writing?
3. Are sentences well structured?
    a. Are there any thoughts, sentences, or paragraphs that you have trouble extracting meaning from?
4. Does the essay flow?
    a. Can it be followed? Are there transitions that make sense?
    b. Are there sections or sentences that are disorienting?
5. Is word choice simple and clear?
    a. Does the author use multiple words when fewer would be clearer?

5 – Topic Essay Exercise

This essay will provide your reader with an introduction to the topic you are choosing to develop. This statement should provide a definitive description of the topic within the research area. The statement should answer: why is the topic important in the research area (or why does it matter, or so what?!)? It should briefly contextualize the topic into related, similar previous research. Further, it should also introduce several research questions that relate to the topic that can be used to either solve practical problems or to increase understanding related to conceptual problems. At the end of the essay, you must choose a research question that you will develop further and state why it is of particular interest to justify your choice. You must also comment on what aspects of peer review from your group that did and did not help you. This assignment corresponds to subject matter presented in Chapter 3. The exercises here also relate to the Peer Review Exercise presented in Chapter 4.

Page limits: Minimum 2 full pages; maximum 4 full pages.

Reviewing the Topic Essay:

Content Review

1. What is the topic the author has chosen?
2. Is it clearly situated into the area of research interest?
3. Is it clear why the topic fits within the area of interest?
4. Is the topic clearly described; that is, does the author state what it is?
    a. Is the topic too broad or narrow?
    b. What words are used to express actions or relationships that focus the topic?
5. Is the importance of the topic reflected upon in such a manner that it becomes clear to the reader?
6. Are three to four clear research questions related to the topic introduced?
    a. Is it clear how these questions link to the topic?
7. Has the author chosen which question they will develop and then stated why?

Readership

1. Are key concepts defined?
2. Does the author cite sources that increase your confidence in their writing?
3. Are sentences well structured?
    a. Are there any thoughts, sentences, or paragraphs that you have trouble extracting meaning from?

4. Does the essay flow?
    a. Can it be followed? Are there transitions that make sense?
    b. Are there sections or sentences that are disorienting?
5. Is word choice simple and clear?
    a. Does the author use multiple words when fewer would be clearer?

6 – Degree Planning Exercise

There are three resources for mapping out your pathway to a master's degree. First, you must read the Student Handbook (I am including ours from UNT Geography as Appendix 2, but seek out relevant documents for your own program) to become aware of program expectations about major milestones and their timing. Second, you must become aware of important deadlines for milestones, including graduation application and thesis submission, in your department, college, and graduate school. Finally, once informed about the first two, you must bring that information to your major professor to agree upon reasonable personal targets for timing of proposal and thesis outcomes. Your plan should have a detailed timeline of expectations related to milestones, including those from your department, college, and graduate school. In addition, you need to incorporate estimates for lengths of time for data collection, analysis, and writing. Submit an email from your major professor that they have met with you about the plan and provided feedback before it is completed (before the due date). This assignment corresponds to material presented in Chapter 2.

7 – Question Essay Exercise

This essay will provide your reader with an introduction to the question you are choosing to address in order to focus on a research problem. In this essay, you should be narrowing down to precisely what your research will cover. It should briefly contextualize the question into the topic, research area, and literature on similar, previous research. Your essay should also introduce what types of claims you hope to make, the reasons that link to those claims, and how you plan to assemble evidence. What will your data sources be? If you are at the proposal stage, outline the tasks, time allotment for tasks, and target dates for completion of the proposal. If you are at the thesis-writing or report-writing stage, do this for a chapter or research report. This assignment corresponds to subject matter presented in Chapter 5. The exercises here also relate to the Peer Review Exercise presented in Chapter 4.

You must also comment on what aspects of peer review from your group did and did not help you.

Page limits: Minimum 2 full pages; maximum 4 full pages.

Reviewing the Topic Essay:

Content Review

1. What are the components of the assignment from the instructions?
2. What is the question the author has chosen?
3. Is it clearly situated into the topic of interest?
    a. Is it clear why the topic fits within the area of interest?
    b. Is a context for the research presented?
4. What is the condition that relates to the question?
    a. What are some consequences of the question not being answered?
    b. If the question is answered, what types of claims could be made?
    c. What types of evidence will be required to make such claims?
5. Does the author identify and discuss important data sources for their research?
6. Does the essay finish with identification of tasks, time allotment, and a plan with target dates for completion? Is it a solid plan or does it seem too vague and loose?

Readership

1. Are key concepts defined?
2. Does the author cite sources that increase your confidence in their writing?
3. Are sentences well structured?
    a. Are there any thoughts, sentences, or paragraphs that you have trouble extracting meaning from?

4. Does the essay flow?
    a. Can it be followed? Are there transitions that make sense?
    b. Are there sections or sentences that are disorienting?
5. Is word choice simple and clear?
    a. Does the author use multiple words when fewer would be clearer?

8 – Personal Research Profile

After completing the set of exercises in Appendix 1 and related ones in Chapters 1 through 5, it is important to synthesize one's research interests. As students become researchers, they are increasingly in professional and social contexts where it is important to summarize their research interests. This exercise provides prompts for developing one's comprehensive research profile. This exercise provides an important transition from the essay exercises—triangulating from interest to topics to questions—toward the literature review (Chapter 6) and proposal storyboard (Chapter 7 and Appendix 3). The prompts in this exercise are drawn from ones provided by Booth et al. 2008, Chapters 3, 4, and 5 at times duplicating their approach. Students should respond to each prompt below in the form of short essays.

1. Describe your area of research interest.
2. List five keywords that highlight your area of research interest.
3. What topic are you either developing for your research or leaning toward? State the topic as a sentence.
    a. I am studying…
    b. …because I want to find out…
    c. …in order to better understand…
4. List additional keywords (up to five) that characterize your topic.
5. Does your research confront a practical or conceptual problem?
    a. What is the condition of the problem?
    b. What is the consequence/cost of the problem?
6. What are some types of data that will be needed to address your research problem?
7. What literature will you need to engage in order to provide a context for your research problem?
8. In order to start a bibliographic trail, what are five critical sources that you can start with?
9. What objectives will need to be met in terms of data acquisition and analysis to address your problem?
10. Are there any skills you will need to refine or acquire to meet your objectives?
11. Briefly describe time allocation and time commitments required to achieve the objectives.

Appendix 2

Student Handbook – UNT Master's of Science in Geography

This handbook is designed to give students a description of the graduate program in geography at UNT. I provide it here as an example related to exercises presented in the chapters of the book and in Appendix 1.

Students should read the entire document carefully; it contains important information on what to expect during the program.

Coursework

The Master's of Science degree with a major in geography has a minimum requirement of 36 hours of academic credit, which includes a 6-hour thesis course.

Students are required to take core classes within the first three semesters:

- GEOG 5160 - Foundations of Geographic Thought Explores epistemological developments in the discipline of geography, including the origins, development and diffusion of predominant ideas that form the foundation of geography. Provides a grounding in contemporary geographic thought, focusing on diverse ways that geographers go about explaining, interpreting and understanding the world (i.e., epistemologies).
- GEOG 5800 - Research Design and Geographic Applications Helps students learn systematic research tools, such as developing a literature review, framing research questions, and writing a research proposal. The class helps students focus their research agenda and develop their thesis proposal. The course provides a structure and format for developing a research agenda, but the proposal-writing process also requires discussion, deliberation, and primary mentorship from the student's major professor.
- Analytical Methods: There is more than one avenue for meeting the analytical methods requirement. Students must take at least one of the following: fundamentals of statistical research (GEOG 5185), spatial and multivariate statistics (GEOG 5190), and qualitative methods (ANTH 5031).
    - GEOG 5190 – Advanced Quantitative Techniques Application of advanced statistical procedures, including spatial statistics and multivariate techniques to analysis of point and areal patterns and spatial data.
    - GEOG 5185 – Statistical Research in Geography Application of fundamental statistical techniques to research in geography, emphasizing the construction of research papers and proposals.
    - ANTH 5031 – Ethnographic and Qualitative Methods Focus on ethnographic and qualitative methods and development of skills necessary for the practice

of anthropology. Emphasis is given to qualitative approaches to data collection and analysis, grant writing, and the ethics of conducting qualitative research.

A replacement analytical methods course can be recommended by the major professor and thesis committee; however, such a course must meet goals similar to those listed above. Exception requests will be deliberated by the Graduate Committee (GC) and will not count toward the requirement without approval of the GC.

The rest of the coursework in the MS Geography curriculum consists of a series of electives that are chosen in consultation with the major professor and the thesis committee (see next section).

Milestones

The goal of the Master's in Science in Geography program is for students to develop a foundation in in the discipline of geography and acquire skills that will allow them to become independent researchers and project managers. Four milestones mark progress toward this goal, all of which are achieved in collaboration with the major professor: 1) formalize a degree plan; 2) structure a thesis committee; 3) write, present, and defend a thesis research proposal; and 4) conduct thesis research and write, present, and defend a thesis.

The Degree Plan

The degree plan is a strategic plan for student coursework and the research hours they will need to acquire the skills and perspectives they would like to gain from the graduate program. Course selection relates to the goals of the thesis research, and thus the degree plan is developed during the first semester through discussion with the major professor. Individual student degree plans and the composition of the thesis committee are defined by the beginning of the second term/semester of attendance.

The Thesis Committee

During the period in which the major professor and student discuss the degree plan, conversations should also develop concerning planned research and organization of the thesis committee. Through discussion, the major professor and the student will deliberate about which faculty members fit the goals of the developing research and can provide additional guidance. The thesis committee is chaired by the major professor and consists of at least two additional faculty members, one of whom must be a faculty member in the Department of Geography. Thesis committee members from outside UNT must be approved by the Department and Toulouse Graduate school; for outside members, the major professor and student are responsible for soliciting a CV and producing a brief statement (via email to the Graduate Program Coordinator) as to how the external committee member is appropriate for the research project.

The Thesis Proposal

All students should defend a thesis proposal by the end of the second semester. Research topics and objectives should be defined and honed during the first semester of the program in GEOG 5800; that course provides a structure and format for developing a research agenda, but the proposal should be developed through discussion and consultation with the major professor during the first and second semester. The proposal consists of a written document that provides a literature review to contextualize the proposed research, a statement of objectives, and a plan for how the research will proceed. The proposal is constructed and written with the guidance of the major professor who critiques and edits drafts of the document. Upon approval by the major professor, a draft of the proposal is sent to the remaining members of the thesis committee, and if they deem the researcher ready, a proposal defense will be scheduled. The defense is a presentation of the proposal. It is open to faculty members and students. The presentation is advertised through the Department, and students must provide the administrative assistants with the date, time, and location of the proposal defense. After general discussion from the audience, the thesis committee will meet with the student to offer feedback on the proposed research. The committee will then deliberate to determine whether or not the student passes the proposal defense, if revisions are required, or if the student cannot proceed in the program. Students who do not pass the proposal defense may consider a non-thesis pathway to the master's degree upon recommendation of the thesis committee and the Department. Students who pass their thesis proposal defense may proceed in their research.

The Thesis Defense

The master's thesis is the pathway to becoming an independent researcher who can develop/address research problems, conceptualize solutions, generate appropriate data, provide analytical strategies to summarize and make inferences based on those data, and synthesize the importance and meaning of research through writing and presentation. The thesis is constructed and written with the guidance of the major professor who critiques and edits drafts of the document. After approval of the major professor, a draft of the thesis is sent to the remaining members of the thesis committee, and if they deem the researcher ready a thesis defense will be scheduled. The defense is a presentation of the thesis, which is open to faculty members and students. The presentation is advertised through the Department, and students must provide the administrative assistants with the date, time, and location of the thesis defense. After general discussion from the audience, the thesis committee will meet with the student to offer feedback on the research. The committee will then deliberate to determine whether or not the student passes their thesis defense, if there are revisions required, or if the student's work does not warrant completion of the program. Students who do not pass the thesis defense may consider a non-thesis pathway to the master's degree upon recommendation of the thesis committee and the Department. Students who pass their thesis defense must submit the thesis to the graduate school under their guidelines and deadlines to graduate. Thesis guidelines may be found here: http://tsgs.unt.edu/new-current-students/theses-and-dissertations. Graduation information may be found here: http://tsgs.unt.edu/new-current-students/graduation-information.

Conference Travel

Faculty members and graduate students regularly travel to a variety of regional, national, and international research conferences to present research findings and to create and maintain professional networks. Students may apply for travel assistance through the Toulouse Graduate School through their Travel Grant Program: http://tsgs.unt.edu/new-current-students/travel-grants.

Teaching Assistantships

Applicants to the program are considered for funding through a teaching assistantship. Generally, there are three types of assistantships: 1) TAs teach lab sections of Earth Science (GEOG 1710), Physical Geology (GEOL 1610), or Archaeological Science (ARCH 2800), which entails a load of four lab sections per semester; 2) TAs assist in teaching specialized courses, such as Foundations of Geographic Research (GEOG 2110), Statistical Research in Geography (GEOG 4185/5185), Advanced Quantitative Techniques (GEOG 5190), or Introduction to GIS (GEOG 3500); or 3) TAs serve as tutors and graders dedicated to multiple lecture and lab sections of Earth Science and/or Physical Geology. Students can read about these courses in the current UNT Undergraduate and Graduate Course Catalogs. *During the application process, if selected for a teaching assistantship, the GC will offer a funding commitment to the applicant for the coming academic year, or potentially for two years. This offer is based on a ranking of all graduate applications. Funding may not be renewed for the second year if a student is not making satisfactory progress toward degree milestones by the end of the second semester.* Typically, students who receive two years of funding, whether from an initial two-year commitment or from two one-year commitments, are unlikely to receive additional semesters of support. If a student's program extends beyond two years, the student can request to be considered in the TA selection pool. New applicants will be prioritized during the selection process over those who have received two years of support.

If a TA is unsuccessful in the classroom, the student can lose funding during or after the semester working with the Department (removal during the semester is exceptionally rare). Significant instances of poor quality teaching can lead to removal between semesters despite a one or two year commitment from the GC. If a student receives a research assistantship during her/his program, the student is still expected to complete the MS degree in the two-year window. The student may or may not receive TA support after two years in the program and will not be prioritized over new applicants or students who have only received one-year commitments in previous years. In rare instances, the Department will have additional TA positions and may approach former TAs who have been in the program for more than two years with support.

Annual Review Process

Note: annual reviews are for addressing relatively normal issues with TAs or graduate students. Serious concerns about misconduct, health, or any other important immediate concern will be under the purview of the Department Chair.

The Monday following Spring Break each year, the GC Chair will send an email to all faculty members requesting feedback on whether students are struggling to meet program goals and/or if they are struggling as a TA. For example, a student may be delayed on one aspect of her/his program (e.g., completion of the degree plan or finalizing of the proposal), but could otherwise be making excellent progress. In contrast, a student might be meeting program expectations and being doing poorly in some areas, such as her/his teaching duties. Faculty members will respond with any concerns for a particular student by April 15. The Chair of the GC will facilitate a meeting with the student and the major professor and/or instructors for whom the student TA'd to seek to resolve any problems. Students who are doing poorly may not receive the same level of support during their second year. It will be assumed that no emails about a student means that major professors, TA supervisors, or instructors are happy with that student's progress.

Additional Information

No grade below a B will count toward the degree. Any grade below a B must be replaced by retaking the course and earning at least a B. Students may retake no more than two such courses. A third grade below a B will result in the student being dismissed from the program.

At the completion of 30 semester credit hours, students will not be allowed to change their initial decision to choose either the thesis or non-thesis option.

Graduate students who have not graduated within one year after completion of coursework must formally apply for an extension to remain in the program (see www.geog.unt.edu for details). If a student does not demonstrate satisfactory progress toward completion of the thesis or research report within 1.5 years of successfully defending the thesis proposal, a grade of F will be automatically assigned for the thesis. Students have the right to appeal this decision to the graduate committee.

Students may elect to follow one of the specific degree tracks currently offered: applied geomorphology, environmental archaeology, urban environments, water resources management, applied GIS, business geography or medical geography.

Tracks

The following tracks serves as guides for those who wish to have a more focused program. These tracks reflect some of the research areas of faculty members in the department. Students are not required to choose a track.

Globalization, Cities and Development

Our global society is more interconnected and interdependent than ever before. Globalization of trade and commerce has increased national wealth and our appetite to consume commodities, technologies, art and culture from around the world. We continue to create spectacular cities to represent our cultural, technological and architectural achievements. But even as we continue to generate extraordinary wealth, we live in a world that is riddled with social and environmental unsustainability, poverty, inequality, discrimination, prejudice, marginalization, terror and conflict. The objective of this track is to train students to understand the complexities of our global society, our cities and our unequal geographies of life and livelihood. Upon graduating, students will find themselves well trained to pursue doctoral degrees, or careers in government, think tanks, non-governmental organizations, teaching, diplomacy and elsewhere.

Applied Geomorphology

Applied geomorphology emphasizes geomorphological processes that are of societal significance, including hazards such as flooding, expansive soils, landslides and coastal erosion. This track enables students to structure their degree plans around conceptual and technical aspects of applied geomorphology. The track meets all existing requirements for the degree, including required courses in research design, quantitative techniques, and a cognate field. Students completing this track may find employment with government research and regulatory agencies, municipalities, planning organizations, water supply districts, or environmental consulting firms.

Environmental Archaeology

Archaeology faculty in the geography department, in cooperation with the graduate program in anthropology, direct graduate students in pursuit of either the MS in geography or the MS in interdisciplinary studies. The focus of this program is to give students a strong foundation in selected areas of research that will prepare them for entry into research positions or doctoral programs in archaeology. Two principal areas of training are geoarchaeology and zooarchaeology, which derive strength from the faculty and laboratory/collections resources at UNT. In addition to core requirements in geoarchaeology or zooarchaeology, students complete two areas of specialization selected from the following areas: GIS and remote sensing, spatial and quantitative analysis, instrumental techniques (e.g., SEM, EDX, PIXE, stable isotopes, petrography), or zoology and ecology.

Urban Environments

This track prepares students to assume a vital role within the structure of a city government, coordinating the activities of various city departments related to environmental legislation. In addition to the normal requirements, students select courses from content areas, including urban environments, environmental science, city government structure, and environmental law and policy. This track has been developed in response to the increasing need for persons to coordinate different programs in city government, to liaison with governmental agencies, to interact with contracted environmental engineers and to bring a philosophy of sustainable environments to the planning process.

Water Resources Management

This track prepares geography students to assume active roles in addressing the critical issues of water supplies and water quality. Students follow a curriculum balanced among technical, scientific and political aspects of water resources management. Courses are selected from the following topical areas: techniques, geography/geology, environmental science and environmental policy. Students completing this degree track gain positions with local and regional governments, federal and state regulatory agencies, engineering firms and regional water districts.

Applied Geographic Information Systems

This track prepares students to meet the growing demand for GIS professionals. Rather than a strictly technical preparation, students acquire the foundation in applied geography that qualifies them to play vital roles in planning, policy and implementation in chosen areas such as urban geography, economic/business development, environmental science and medical geography. Courses for this track are selected from a chosen subset of the following groups: GIS technology, GIS applications, topics/cognate fields, real estate/marketing, public health administration, environmental science and applied economics.

Business Geography

The objective of this track is to educate students to integrate geographic analysis, reasoning, and technology in support of improved business decisions. The focus on improving the decisions made by business differentiates business geography from urban/economic geography. Participation in a business internship is encouraged. If appropriate, the results of the internship can form the basis for the student's MS thesis or problems in lieu of thesis.

Medical Geography

This track focuses on theory and techniques that are needed to understand the spatial patterns of health outcomes, environmental risks and exposures and disease spread, as well as the distribution of health care services and lack thereof. Students specializing in this track will learn about the relationships between human activities, place, and health outcomes and how to evaluate those relationships using GIS methods, spatial and statistical analysis, and computational models.

Appendix 3 – Research Product Assignments

I separate these exercises from those in Appendix 1 because these are designed to develop the thesis proposal, which is a research product linked directly to an important milestone of the master's degree.

There are three assignments that cover your research product for the semester, each of which has evolved and should continue to evolve during the next month as you refine your topic, settle into research questions, and develop your literature review (Chapter 6). It is not necessarily expected that you finish the semester with a complete proposal. However, aspects of your product should be coming together this semester; thus, these exercises are actually progress reports in which you show awareness not only of what it is that is coming together already, but also tasks, target dates, and time requirements for components of your research that are still in development. As a result, there should be some parts of your storyboard, your presentation, and proposal (or report or chapter) that are more developed than others; use this as an opportunity to assess and strategize.

Proposal Storyboard: For this exercise, I am looking for you to use the storyboard template and to flesh out areas in which you are farther along, but also to help assess and make plans in areas that need development. This will be an outline with some areas that are more detailed and others that represent planning to accomplish tasks. Follow the storyboard template, and use prose to flesh out sections as possible. This exercise relates to material presented in Chapter 7; a template is provided in the next section of this appendix. Booth et al. (2008: 175–176) provides a helpful introduction to storyboards. The storyboard should evolve from outline to proposal narrative over time.

Research Presentation: For this exercise, you will convert your storyboard into an oral presentation in which you communicate aspects of your research that are developed. Given your work this semester on essays and the annotated bibliography, the context for your research should be fleshing out, and you should be identifying questions and their conditions and consequences. This is an opportunity to engage the class members as an audience, to share how your research is coming together. Given that your research is ongoing, the presentation should represent some progress beyond the initial storyboard assignment. This exercise relates to material presented in Chapter 8; a template is provided later in this appendix.

Proposal Draft: During the final month of the semester, our readings concentrate on how to become a research writer, one whose habits lead to research outputs. This stems directly from the work we did early in the semester to determine research foci (the essays). Lessons from Booth et al. (2008), and from additional readings related to editing, proofreading, and revising communicate how to develop this important research skill. In addition, the readings by

Silvia provide advice and support on how to manage the writing process. Your product will, once again, derive from the storyboard from the previous exercise. In this exercise, you will write the portions of your research that are developed, and you will write about those components of research that still need development in your work. The result will be part product and part plan, all written as narrative. As with the presentation, it is expected that some aspects of your research will evolve during the final weeks of the semester; thus, some components of your research should be farther along in the report than in the storyboard and presentation.

Proposal Storyboard

What sections and subsections will be required to complete your proposal?

Which sections are you prepared to outline currently?

What are important target dates for completion of the proposal?

What are important non-writing responsibilities (e.g., meetings with major professor) that need to be planned?

Proposal Introduction (roughly 2 to 3 pages)

The Topic and Research Question
Note: you will eventually write the introduction by establishing common ground with your reader to engage their interest, but start with simply knowing where the proposed research is headed.

The topic to be studied (a subject combined with action/relationship words):

The topic focused into a research question (to find out what/why/how?):

Turning the question into a problem by gauging the significance of answering the research question (to understand...?):

Importance of the Research Problem
The condition/situation that answering the research question addresses:

The consequences of not answering the research question:

Purpose of the Study
Note: this will not be a full statement of your objectives, but it will need to be a concise and clear statement that orients the reader prior to diving into the literature review, which will provide the fuller context of your research.

What the study will do in order to address the research problem (what you intend to study/do in the research):

Additional notes on the introduction:

Literature Review (roughly 3 to 5 pages)
What research area does the topic fall under?

Context: is there a particular social, environmental, political, or other type of context that is of particular importance related to the research problem?

What concepts must be defined in order to proceed in the research?

What previous studies address the same or similar topics in the research area? Document any historical trajectory of research related to your research problem.

Do any previous studies address your research problem, perhaps in a different place, using a different methodology, or both?

Summary of research methods commonly used to address similar research problems.

Outcomes (what was learned: results, conclusions, recommendations) from previous studies.

Any other notes on the literature review:

Study Objectives (roughly 1 page)
Restate the research question and problem, its condition and consequences:

Itemize three to four objectives that must be accomplished to address the research problem:

Other notes on the objectives:

Proposed Research Methods (roughly 3 pages)
Note: these steps are the same for any type of study, ranging from environmental archaeology to geomorphology to political ecology, using qualitative, quantitative, or mixed methods.

Study Area—the details of where the study will take place:

Data requirements—the details of what data are required to address the research problem:

Data collection—the details of the tasks required to assemble data:

Data organization—the details of how data will need to be manipulated to prepare for analysis:

Analytical Methods—the details of how data will be examined and/or analyzed to address the research problem:

Other notes on methods:

Research Plan (roughly 2 pages)

Tasks and Sub-tasks—assemble and organize the action items needed to meet the objectives of the study:

Time Allocation—assemble time estimates for completion of tasks and sub-tasks:

Timeline—provide dates for completion of tasks and sub-tasks:

Other notes on research plan:

References Cited

Note: we have used the APA style in this class, but you may need to adjust to what is expected in your research area for your actual proposal. Come to an agreement with your major professor about which journal's style guide should be used for your research, and then stick to it all of the way through your thesis. Continue to use APA style in this class.

Provide a formatted complete list of the references cited used in your proposal in alphabetical order.

Proposal Template

Steve Wolverton

Introduction

- Use a relevant image; don't get wordy (use index cards)
- State your topic and what field it is in
- Road map for presentation

Context

- Don’t get wordy; pick an image from a relevant case study
- Brief, general use of previous literature
- Identify the gap that your research fills
- Identify and define key concepts

Research question

- State the research question(s)
- Articulate the consequences of not answering the question(s)

Objectives

- Articulate 2 to 3 precise objectives that the research will meet to address the research question(s)
- These can start to point the audience toward your study area and methods

Study area

- If relevant, describe where your research will take place
- State why the study area works for the research

Proposed methods

- Report the status of what you know about methods
  - If you know exactly what to do, then report that
  - If you are in the process of learning methods, report what previous studies have done
    - If you need to learn, make sure that is clear
- What are the data requirements?
  - How will data be collected & organized?
- How will data be analyzed?

Preliminary Results

- Report any progress that has been made

Research plan

- Tasks and sub-tasks
- Time allocation

Conclusion

- Restate
    - Questions & consequences
    - Objectives
    - Basic plan

Made in the USA
Middletown, DE
27 March 2020